Post-industrial Robotics

Angelo Figliola · Alessandra Battisti

Post-industrial Robotics

Exploring Informed Architecture

Springer

Angelo Figliola
Planning, D&T of Architecture
Sapienza University of Rome
Rome, Italy

Alessandra Battisti
Planning, D&T of Architecture
Sapienza University of Rome
Rome, Italy

ISBN 978-981-15-5277-9 ISBN 978-981-15-5278-6 (eBook)
https://doi.org/10.1007/978-981-15-5278-6

This Springer imprint is published by the registered company Springer Nature Singapore Pte Ltd.
The registered company address is: 152 Beach Road, #21-01/04 Gateway East, Singapore 189721, Singapore

Preface

This book highlights the concept of informed architecture as an alternative or implementation of a performance-based approach. Starting from the analysis of the state of art, the book will define an operative methodology through which performative parameters lead and inform the generation of the shape becoming design's input rather than mere quantitative parameters. Visually dynamic responsiveness, obtained by the morphological reconfiguration of the systems, become static through informed design proceedings. Within this scenario, generative design and digital computing play a key role in the efficient exploration of design solutions, as well as for the ability to focus formal generation, simulation of dynamic phenomena, and manufacturing on a single workflow. The methodology will be investigated through the analysis of case studies. The project's analysis aims to identify the basic research lines, showing the main results obtained and extracting the theoretical concepts that follow the investigations carried out at universities and research centers active on the matter. Specifically, the research lines have different approaches to the topic in relation to the typology and nature of the relationship between academic studies and industrial procedures; the use of traditional and experimental materials and production processes; the exploration of innovative concepts linked to collaboration between man and machine and cyber-physical making. In addition to a different way of conceiving and utilizing the manufacturing tool, at the base of the survey there is a different interpretation of the performance concept, which is the basis for the definition of informed architectures in relation to data usage and the optimization procedures. This book presents results of the studies titled *Post-industrial Robotics* conducted from December 2014 to August 2016 under the supervision of Prof. Alessandra Battisti at University of Rome La Sapienza Doctoral Program.

Rome, Italy
February 2019

Angelo Figliola

Contents

Chapter 1
Exploring Informed Architectures

Abstract The setup of the first robotic manufacturing laboratory in 2005 at the Swiss Federal Institute of Technology Zurich (ETHZ), marked a new course for digital manufacturing in architecture by introducing innovative design paradigms that soon became the cornerstone of major researches on the topics. A generic industrial machine is a tool used to transform virtual models into material systems with direct and bidirectional connections between the digital model and the production process. The potentialities of this methodology are investigated over several analyzes aimed at designing and fabricating performative and informed architectures through a digital protocol that includes computational design and morphogenetic, material computation, and innovative manufacturing methods. In that respect, the purpose of the chapter is to identify the main study lines, showing the results obtained and extracting the theoretical concepts resulting from investigations carried out by universities and research centers invested in the cases of informed architectures.

Keywords Informed architectures · Parametric design · And digital fabrication · Post-digital · Performative architectures · Material culture · Research paths

1.1 Introduction

The recent publication *Parametricism 2.0* [51] has marked the beginning of a new phase for the post-digital age aimed for the use of computational, algorithmic tools for the resolution of environmental and social matters, readdressing issues that determined the birth of digital elaboration itself in the '70s [14]. The ability to process information and later utilize the data as guiding elements of the design evolution [13] opens many, and largely unexplored, possibilities for environmental and technological conception. The theoretical assumption defines the architectural shape as a result of a diagram of forces within a morphogenetic process [23] and opens innovative investigation fields related to the possibility of finding a new kind of performance-based [20, 26, 27, 43], or performance-oriented architecture [24]. So far, the first digital age has interpreted performance as a necessary antidote against formal arbitrariness generated by digital processes, or as a representation of new

A. Figliola and A. Battisti, *Post-industrial Robotics*,
https://doi.org/10.1007/978-981-15-5278-6_1

hyperfunctional complexity. Today, it represents an absolute necessity, an ethical obligation of the profession due to codes and regulations. Operating within this scenario, the technological innovations of the modern era are essential to ensure that performances do not remain only as a quantitative parameter but can be interpreted as a source of formal exploration and convert information. Hence, the generative process has the objective of achieving standard performatives that can be optimized for a space of design possibilities [34] defined by the designer himself through a meta-project setup [28]. Following this approach, designers define a specific goal that can be reached by manipulating geometric variables and constraints. The production progress, through computational design, offers the opportunity to explore complex and informed geometries in a flexible and fast way, transforming the material from a passive recipient of shape to a design agent, as driving element of the creative process, through the exploration of its mechanical, structural and behavioral features [32]. The application of such methodology involves overcoming the typological paradigm and turning to a continuous formal variation that changes depending on the conditions around the system. Thanks to the data-driven strategy and the ability to interconnect conception and manufacturing in a single workflow [52], the customization of the form can be linked to a responsive interpretation of local and regional variations in characteristics [56]. To allow the transfer of an informed architecture from the digital to the physical world, the latter must integrate various skills and shift technologies employed in other areas of architecture to implement standard processes. The birth of the first robotic production plant at the ETHZ in 2005, with the direction of Fabio Gramazio and Matthias Kohler, marked a course for digital fabrication in architecture by introducing new design paradigms. The generic industrial machine becomes a tool through which to transform virtual models into material systems with direct and bidirectional connection between the digital reproduction and the manufacturing procedure. The potentialities of this design methodology are explored over several research trajectories which aimed at planning and fabricating performative and informed architectures through a digital protocol including computational design and morphogenetic, material calculation, and innovative production processes. The digital-material relationship allows combining the analysis of novel formal codes with anaphoric aspects that ensure the manufacturability of what is generated in the digital space and applied it to the scale of architecture [16, 55]. The integration of informed engineering and digital constructing technologies opens an era based on the customization of performative architecture. Other disciplines contaminate the hybrid space of interaction between designer and machine in order to survey innovative fabrication methods, but also to stimulate creativity through a fruitful collaboration process [53]. The interdisciplinary collaboration opens a post-industrial phase in which issues related to analog design and manufacturing methods are interrelated with disruptive digital technologies.

1.2 Analysis of the Main Research Lines: Definition of the Theoretical and Instrumental Apparatus

The first chapter proposes a critical summary of the main international research trajectories on the subject of informed architectures and advanced manufacturing technologies (Fig. 1.1). Within the broad spectrum of the topics, the characterization of the primary study paths was developed by identifying the variable and invariant aspects and the methodological and instrumental apparatus that enable the determination of different interpretations on the theme as well as its weak points (Fig. 1.2). This text focuses on design processes that can be traced back to the concept of informed architecture, as a different interpretation of the notion of responsive and adaptive construction. If the concept of responsiveness refers to the ability of the systems to react dynamically to internal and external stimuli, the adjustable term introduces a different layer of complexity, as it combines responsiveness and adaptation for diverse morphological configurations, or programmatic functions, generated by system solicitation. The definition of informed architecture has been examined, focusing the attention on the early stage of the design process and the employment of performative parameters as a formal generation engine that transforms the latter into design inputs. This was done in order to describe the correlations between the operative methodologies—which allows the conversion of theories into design actions, and the use of responsive or adaptive systems at the base of kinetic architecture—and the aspects related to the material computation. Regarding the classification introduced by David Leatherbarrow [30] which defines the responsiveness as visually static or dynamic,

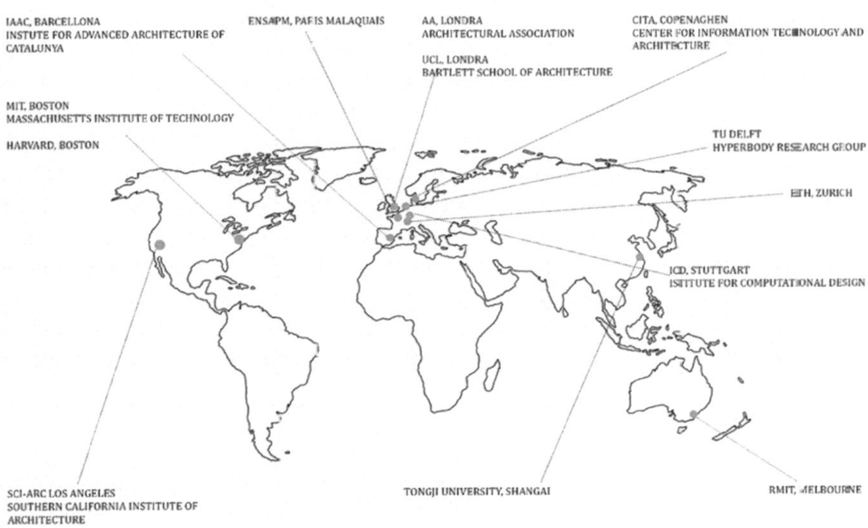

Fig. 1.1 *Post-industrial Robotics*: main architectural schools active on the subjects. © Angelo Figliola

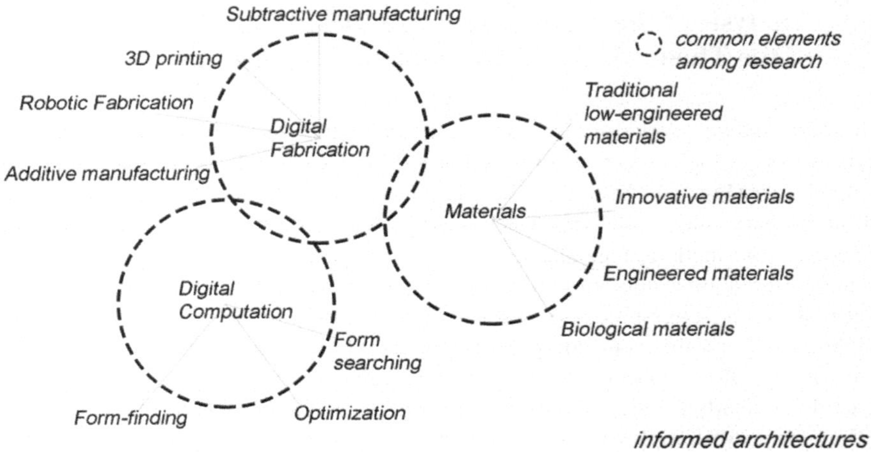

Fig. 1.2 Variable and invariant elements of the analyzed design methodologies © Angelo Figliola

the information of the conception evolution through performative parameters allows the realization of visually static responsive morphologies that can be implemented using dynamic and reactive systems at the local scale. The above-defined data-driven method represents a common thread within the study lines analyzed. Another invariable element is represented by the tools used in the whole design process: digital calculation and computational thinking[1] are the fundamental aspects of experimentation conducted worldwide. The application of parametric designing tools, using visual scripting techniques, allows the structuring of the conception process following a hierarchy of data based on the association and interrelation of parameters rather than on the addition of unstructured information. Furthermore, parametric tools simulate and analyze the action of forces which, in the natural world, guide the formal generation progress avoiding the construction of complex analogical and time-consuming models. If we consider architecture as an element of mediation between internal and external complex systems, simulation becomes an important instrument in the quantitative analysis of performance parameters as well as a provider of instrumental support to the designer in the implementation of the creative process. Compared to traditional performance-based design, digital computation enables the connection of geometric data to quantitative values resulting from analyzes aimed at guiding and assisting the formal generation course. More specifically, the innovative element

[1]Professor Jeannette Wing, director of the Department of Computer Science at Carnegie Mellon University, defines computational thinking as the mental process underlying the formulation of problems and their solution so that the former is represented in a form that can be implemented in a manner effective by an information processor, be it human or artificial. It represents the effort that an individual performs to provide another individual or machine with the instructions necessary to complete the given task.

consists in the direct connection between the design process and production evolution, between design, material, and manufacturing, achievable thanks to computational process information, which increases the tectonic relations between structure and material within the limits of the digital manufacturing logics adopted. Digital manufacturing technologies, specifically robotic production and 3D printing, constitute a further invariant aspect, as means available to the designer through which the range of style possibilities can be expanded applying different processing strategies and production methods. The construction technologies analyzed—such as subtractive, additive, and hybrid practices—involve the use of traditional and technologically advanced elements. From the relationship between geometry, material, and production evolution, arises a creative process aimed at experimenting with innovative formal codes. Once the invariant elements have been defined, it is necessary to examine how each school or research center characterizes its studies by making specific in-depth analyzes regarding the issues related to digital computation and performance-based approach, materials, and creation processes. Starting from the survey of the performance-based procedure, it is possible to define different types of benchmarks which are utilized to guide the generative process acting on the geometric variables established by the designers. In particular, we can outline two main performative aspects: energy and environmental, and structural conduct. The energy and environmental performance refer to the improvement of indoor comfort conditions and thermal behavior through the optimization of parameters that regulate these aspects, such as daylighting, incident solar radiation, and natural ventilation. As far as the structural aspects are concerned, there are two types of inquiries made: form-finding methods—which provide for the definition of the best spatial configuration for structures subject to compressive and/or tensile stresses, with consequent optimization of the material distribution—and topological optimization processes and distribution of structural supports, as well as the definition of the minimum resistant sections in relation to different load conditions. A further performative layer consists in the informed distribution of the components following stress lines and structural patterns. Compared to digital computation, the main distinction between the various methodologies can be operated considering the optimization progress and the design strategies adopted to perform these actions. The development procedure permits the optimization operating with genetic algorithms or form-finding strategies, both linked to the quantitative analysis of the performances identified as guide elements of the design process. On the subject of genetic optimization, the algorithm turns out to be an important design tool to structure the data of the 3D models in such a way as to identify the variable geometric principles—genotype—whose variation is linked to specific values of the studies conducted—phenotype—to operate the optimization with genetic solvers integrated into the regulation software. On the other hand, the form-searching processes benefit from the flexibility of the parametric models which enable a quick and efficient exploration of different design options in order to identify the optimal solution that integrates formal choices with quantitative performance evaluations. Given the complexity that genetic optimization strategies require in structuring the algorithm and making the calculation, the heuristic form-searching approach prevails, especially in multi-objective optimization proceedings.

Finally, a further variable of the researches consists of the materials employed in the experiments: the integration between computational process and innovative production methods allows the reintroduction of traditional and natural substances, such as clays, stone, and wood in the design evolution, enabling the exploration of the relationship between form customization and optimization of performative parameters. Among the materials listed above, a further distinction must be operated within the area of their origin.

In the case of clays, we can identify the following types:

- Industrial-derived ceramic, composed of clays, feldspars,[2] silica sand,[3] iron,[4] and aluminum oxides[5] and quartz[6];
- Natural clay composes as follows: 90% natural earth, 6% water, 0.8% gelatinous proteins, and 3.2% hexametaphosphate.[7]

The two material types are distinguished in terms of production processes: the fabrication of ceramic components includes a drying and firing procedure using high-temperature ovens. The drying process of the manufactured components is necessary to allow evaporation of the residual water contained in the compound. It can traditionally take up to a week, but it can also be accelerated using special drying tools. Once the water evaporated, the components are placed inside an oven at a temperature varying between 900° and 1100° depending on the different cooking methods. The natural clay does not provide the above-described procedure, and it follows the drying process exclusively. Tests carried out on the substance have shown its excellent mechanical characteristics which allow increasing its structural rigidity as well as its good environmental footprint—considering that the material is biodegradable and directly available on site. As for clays and also for wood, it is possible to make a distinction regarding the level of engineering needed to employ the component in construction methods. In addition to industrial products, such as profiles and laminated panels, advanced digital fabrication technologies enable working directly with raw materials, such as shrub trunks, without resorting to industrial engineering processes. Innovative technologies allow to materialize complex geometries with elements characterized by irregularity in geometry and poor physical and mechanical properties. In addition to the traditional and natural ones, the processes analyzed involve other innovative materials in architecture and engineering, among which thermoplastic polymers, biological elements synthetically produced in a laboratory, and composite materials like fibers and resins. Starting from plastic polymers, plastic derivatives with high molecular weight, represent a new resource for the construction of architectural components. Polymers are filiform molecules of an organic nature, consisting mainly of carbon and hydrogen atoms with the presence of oxygen and

[2]Group of natural minerals which constitutes 60% of the earth's crust.

[3]Sedimentary rock produced from the erosion of other rocks.

[4]FeO, ferrous oxide.

[5]AI2O3, aluminum oxide.

[6]SiO2, Silicon dioxide.

[7](NaPO3)6, Sodium hexametaphosphate.

nitrogen. The polymers most commonly employed in architecture are thermoplastic polymers capable of acquiring malleability through the action of a heat source which makes the material viscous and therefore capable of assuming a great variety of morphological conformations while still being able to return rigid, and thus acquiring stiffness. This feature makes them reusable and re-formable several, theoretically infinite, times ensuring the recyclability of the substance. Acrylonitrile butadiene styrene (ABS) (chemical formula: C8H8 C4H6 C3H3N) n and polylactic acid polymer of lactic acid (PLA) (chemical formula: C3H6O3 CH3–CHOH-COOH) are two of the most common polymers in 3D printing. Fibers, especially glass and carbon ones, belong to the same family of plastic polymers. These, together with synthetic resins, polyamide or epoxy, contribute to the realization of structural composites characterized by a low ductility compared to other plastic polymers, but with a high structural quality through which they can withstand high breaking loads. A further category of materials that can be created thanks to synthetic laboratory processes are biological elements whose physical and mechanical properties can be programed starting from chemical formulas extracted from the analysis of biological organisms available in nature (Fig. 1.1).

1.3 ETH Zurich, A New Relationship Between Architecture and Production

The Swiss Federal Institute of Technology Zurich (ETHZ) has played an important role in the definition of innovative research paths on computational and digital fabrication developed in the most important universities and research centers in these last years. In this scenario, architects and researchers Fabio Gramazio and Matthias Kohler of Gramazio Kohler Research[8] at ETHZ were fundamental in defining, not only a new way of conceiving manufacturing processes based on the power of digital computing,[9] but also for having introduced a series of theoretical concepts that have become a starting point for other studies [18]. The research carried out at the ETHZ has some distinctive characteristics:

- The first aspect consists in the strict collaboration with the industrial world, which transforms research projects into patents and subsequently into technological solutions applied in the construction and manufacturing processes;
- The second aspect concerns the scale of experimentations and planning purposes. Most of the analysis carried by the research group mentioned above, as well as the results of the Master, MAS ETH in Architecture and Digital Fabrication, produce prototypes of technological systems or architectural organisms as 1:1 scale experimental output. Such prototypes can show the critical issues and the potential of

[8]www.gramaziokohler.arch.ethz.ch/web/e/forschung/221.html.

[9]The reference regards the mTable project of 2002, where the potentials of digital fabrication are linked to digital computing and the possibility of customizing shapes in relation to the individual user's wishes.

an innovative design methodology. In collaboration with AEC industries and with a 1:1 scale, they are able to solve the ratio, feasibility, and applicability related problems, which have characterized the digital fabrication studies stated by Carpo [10] paper: "The best, the most persuasive, the most conspicuous results of the new fully integrated seamless file-to-factory technologies have happened on the scale of the teapot".

- The third aspect consists in the ability to experiment by involving the different disciplines and knowledge that define the post-digital age: from software engineering, to material systems, artificial intelligence, and mechatronics, the researches explore all the components that can play an important role taking into consideration the whole system, without focusing on a specific element [17].

The quality of their research and its effective application in the architectural and industrial field led to the birth of the National Center of Competence in Research (NCCR) Digital Fabrication funded by a Swiss National Science Foundation (SNSF) program, of which ETHZ is the leader and coordinator of other Swiss academic and research centers. The purpose of the studies conducted between 2015 and 2018 is the introduction of computational and digital fabrication technologies in architecture and engineering taking into account the whole process, from the designing phase to the construction one. The concept of design-to-production is materialized through an interdisciplinary and multi-scalar approach that sees the cooperation between six different disciplines: architecture, structural engineering, materials science, computers science, robotics, and systems engineering (NCCR 2016). One of the main focuses of the research is on additive processes, in the broadest sense of the term, and on the complex assembly operations of high-performing technological components. The investigated material systems range from traditional elements, such as clay bricks, wood elements, and metal constituents, to the growth of production and manufacturing systems for fluid-dense elements using thermo-polymers; while the computational strategy adopted focuses on the relationship between optimization of structural and substance performance and subsequent materialization of the virtual process and the digital space. The investigation focuses on all the scales of the project: from the development of customized manufacturing systems to the design of optimized technological units and architectural organisms. The aspects surveyed have as guiding principle the optimization of performative parameters and environmental sustainability, linked to the entire life cycle of the project. The various spatial and technological solutions are generated and modified based on such guidelines. What may seem to be a simple technical exercise based on an excessive trust in technology and mechanical instruments, defined as techno-utopianism [49] of the fourth industrial revolution, presents a series of complex and transdisciplinary theoretical and instrumental innovations [11]. The research carried out at ETHZ can be defined by some peculiarities: the consolidation of the relationship with the industrial and production world, the resolution of problems related to the scale of experimentations and research, and the development of the theory established as digital materiality.

1.3.1 *Digital Materiality: A New Materiality Linked to Process Information*

Through the applied studies at ETHZ it was possible to define and outline a new relationship between virtual digital design and physical space, through materials that lead to the definition of a digital materiality theory. The authors themselves define digital materiality in the text Digital Materiality in Architecture (2008): "With the term digital materiality, we designate an emergent transformation in the expression of architecture. We recognize that materiality is increasingly being enriched with digital characteristics, and these characteristics significantly affect the material nature of built architecture. Digital materiality arises through the interaction between digital and material processes during design and construction" [16]. This theoretical approach represents a virtual bridge between the expressions and the architectural representations of the digital age, defined as smooth architecture[10] [26] and blob architecture[11] [31] and the physical world. While during the first digital age the project was formulated in the virtual environment through the expressive power of soft and sinuous forms, the second digital era introduced the concepts of tangibility through the materialization of the virtual space. The distinction between the theory of *Objectile*[12] proposed by Bernard Cache on digital materialization through a non-standard production method [22] and the Digital Materiality one is evident: for the first time the component plays an active role in defining the form, investigating its performance principles through a feedback loop and the continuous interaction between computation and fabrication thanks to a continuous workflow. The digital process is informed by the material characteristics and the constraints derived from the manufacturing method used, which consequently expands the range of design possibilities and introduces a new performative layer. The conjugation of the two seemingly opposite terms digital, representation of the virtual environment, and material, as something tangible and concrete, finds full expression with the computational-algorithmic approach. Using the methodology described above, defined as algorithmic thinking [19], the performative parameters are included in the design process and the determination of the result. The possibility of establishing a dialogue between the digital procedure and the material system brings back in the spotlight the architecture's element aspect, which characterized the work of modern

---

[10]The term *Smooth Architecture* indicates the proliferation of non-Euclidean geometries based on Bezier or Nurbs curves from the end of the '90s.

[11]The term Blob Architecture Greg Lynn identifies a particular type of surfaces defined isomorphic polysurfaces which are a new type of complex geometries also defined as metaball or blob models.

[12]"Objectile", Bernard Cache and Patrick Beaucè, is an architectural firm that took part in the 2002 Non-standard exhibition at the Pompidou Center in Paris directed by Frédéric Migayrou and Zeynep Mennan; the theme refers to the universe of mathematics, and in particular to that branch called "non-standard analysis" which has paved the way for two important areas of research: computer-based calculation focused on algorithms and morphogenesis. The projects shown here are in fact visible and tangible examples of a new form of standardization which, thanks to the new mathematical design techniques, makes it possible to obtain prefabricated elements that are not necessarily all the same.

Fig. 1.3 *Pike Loop Project*, New York. Informed assembly process. © Gramazio Kohler Research, ETH Zurich

movement architects such as Carlo Scarpa, Louis Kahn, and Mies van Ver Rohe. The research carried out on the matter involves different types of innovative, but also traditional, materials commonly utilized in professional practice. Such components can be reinterpreted in light of progressive technologies. The proposed approach was tested through a series of studies on the construction of technological systems made with traditional clay bricks, which led to the birth of R-O-B,[13] a mobile robotic fabrication laboratory [9]: the structural and spatial performances generated by the complex aggregation of small components operated directly on the installation site were tested through the Structural Oscillation projects for the Venice Biennale in 2008 and the Pike Loop in 2009 (public installation in New York) (Fig. 1.3).

As for the clay bricks, other projects have involved the use of discretized wood components and concrete in order to investigate the design possibilities, through the exploration of performative, formal codes offered by this methodological approach. The West Fest Pavilion 2009 (Fig. 1.4) can be seen as a manifesto which explains the potential of this approach as stated by the authors in a 2010 research paper: "The West Fest Pavilion project (2009) explores which criteria of a material system are decisive for architecture and how the correlation of differing requirements can offer new design potential". Through the digital process, it was possible to consider different criteria such as geometric dimensions, structural, and spatial performances for the subsequent materialization with a robotic arm through a layer-by-layer assembly procedure. Starting from a single geometric element, such as wooden profiles, it was possible to customize the design based on structural completion simply by varying the geometric dimension of the wood components and the vertical stacking process.

[13]R-O-B (Mobile Fabrication Unit) is a mobile robot manufacturing unit that was created with the intent of transferring production outside research laboratories and placing it directly on the production site. The unit consists of an industrial robot placed inside a container and can be used for different types of processing.

Fig. 1.4 *The West Fest Pavilion* project. Robotic fabrication and assembly process. © Gramazio Kohler Research, ETH Zurich

1.3.2 Big in Scale: Robotic Fabrication in Architecture and Construction

Another aspect linked to the analysis conducted by Gramazio Kohler Research at the ETHZ is the scalability of the proceedings and their applicability in architecture. The realization of the timber roof structure of the Arch_Tec_Lab building at the ETH campus Hönggerberg is the highest point of the research path on structural maximization through computational design strategies, material systems, and digital fabrication processes culminating in the application of technological complexes in a 1:1 scale [2]. While the milestone conception, which marked the overcoming of the first digital age, and the definition of the theoretical and instrumental contours of the post-digital era, is undoubtedly the realization of the 2006 Gantenbein Vineyard façade,[14] in Flasch, Switzerland. The digital process has enabled the planning and building of a non-standard façade in order to complete a simple reinforced concrete skeleton by the architects Bearth & Deplazes. The frontage mentioned above had to guarantee the execution of the building as a wine cellar in addition to expressing an architectural idea and consequently completing the project. Through the façade's delineation, apart from aspects related to indoor comfort and energy-environmental performance, the architects aimed at conferring architectural plasticity capable of distinguishing the intervention in its application context. Traditional components, such as clay bricks, were implemented for the construction of the technological system, through a complex aggregation whereby the performance parameters previously listed were integrated into the design and construction process (Fig. 1.5). Thanks to the robotic assembly procedure it was possible to position about 2000 clay bricks setting their precise geometric position in the digital model. The placement of the individual bricks was determined based on environmental conduct, such as penetration of sunlight and natural ventilation, and complying with the need to determine a plastic effect based on a 2D texture that informs the digital proceedings,

[14]http://www.gramaziokohler.arch.ethz.ch/web/e/forschung/52.html.

Fig. 1.5 *Gantenbein Vineyard*, customized façade able to generate geometric patterns and visual effects. © Gramazio Kohler Research, ETH Zurich

through the use of a .jpeg image that drives pattern generation. Furthermore, the customization of the rotation of the single bricks demanded a study of a specific type of fixing based on an ultra-strong linking agent; this technique overcame the need to use secondary reinforcement systems present in standard prefabrication. The 400 m^2 façade, consisting of 72 prefabricated elements, was produced in the ETH laboratories, and later transported and assembled on a dry building site. The direct connection between the digital model and the manufacturing method enabled the automation of the processes and the flexibility of the model, enabling its modification up to the last moment before fabrication.

The project described above is the first of a series of researches that have led to the construction of prototypes for architectural applications in a 1:1 scale. The need to scale production processes and methods defined and customized software for managing and simplifying the design of components and promoting access to the professional world. One of the examples is the Brick Design project which led to the creation of a software environment that designers can use to generate the necessary code to inform the manufacturing progress defining the morphology based on the assembly of discrete elements. The Brick Design Software was a prerequisite for the realization of the FlexBrick concept developed from 2008 to 2010. Through this research project, a technological system has been experimented to combine tectonic expression with a building technique and a sustainable material, such as clay. The façade was realized following a study on the software and of the robotic assembly hardware, aimed at the scalability of the process and its application to the 1:1 scale.

The proposed studies demonstrate the will to connect academic research with the AEC sector trying to introduce improvements in the procedures through product innovation. At the base of the experiments, there is an analysis of the traditional fabrication chain to identify potential aspects that can be enriched by process innovation. The concept can be better explained through the ongoing research project "On Site Robotic Construction" which introduces robots in the work environment. The robots are used to perform multiple tasks programed in the virtual space with the possibility of interrelation and feedback with the physical space through sensor and machine learning techniques. The ability of the robot to execute multiple activities with a computer vision of the environment is the main variation introduced after the attempts made in the '80s, especially in Japan [8], where a sole robotic arm was programed to perform a single specific task, without solving all the difficulties related to the on-site interoperability between robots and traditional workforce. The introduction of robots into the construction site presents innovative components: human-machine interaction, receptive capacity of the robot for the workspace thanks to sensors (Fig. 1.6), and the possibility of correcting tolerances and inaccuracies derived from the manual process.

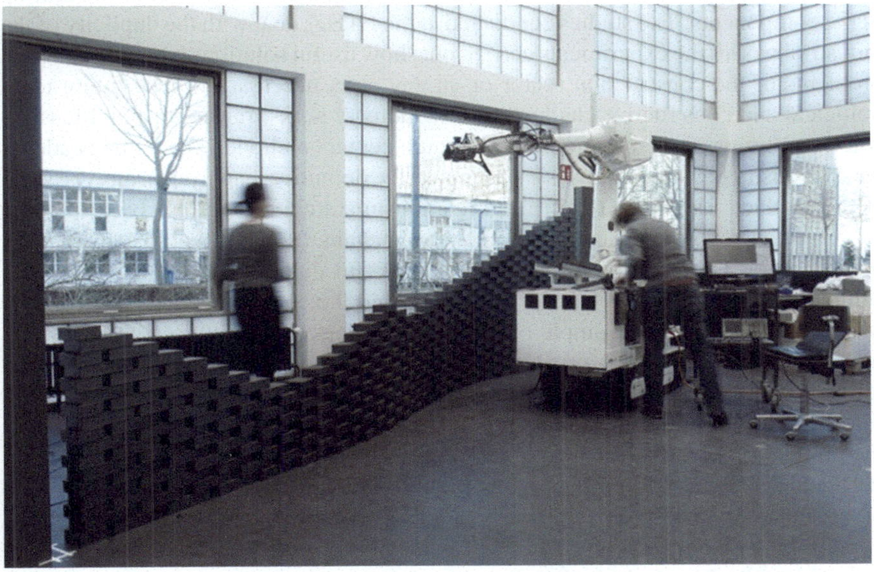

Fig. 1.6 *On-site Robotic Fabrication*: robot's sensitive capacities increase for on-site applications. © Gramazio Kohler Research, ETH Zurich

1.3.3 Architecture Versus Industry: From the Prototype to the Patent, from the Technological Systems to the Tools

The connection between research laboratories and the professional sector through the expandability of processes is directly linked to the relationship established with AEC industries. The studies conducted by Gramazio Kohler Research have, in most cases, collaborative correlations with businesses operating in the construction sector and within this framework they create technological systems as well as tools to be patented and placed on the market in order to innovate a specific technological procedure. From this point of view, the ties with industrial methods turn out to be different from past experiences. During the Bauhaus period, the homonymous school was entrusted with the task of experimenting and innovating through the study and the processing of new and traditional materials and construction systems in order to realize prototypes. The prototypes were then sold to the industry which had the commission of producing the objects in series that were later placed on the market [12]. In this case, the industry was a passive recipient of a creative process and had the sole objective of automating the production procedure for the serial fabrication of the components. The studies conducted at the ETHZ, deal with the duplicity of the industrial world: the experiments carried out show useful solutions based on concrete problems of the construction sector or, in any case, necessary to face future technological and constructive challenges. The industry seeks transferable and feasible solutions through collaboration with research centers and, vice versa, the academy needs industry to validate processes and make the definitive leap from prototype scale to the product. Thanks to this fruitful relationship, architecture goes back to being an important player in the entire manufacturing chain, from concept to construction, incorporating into the digital procedure, thanks to digital computation and algorithmic thinking, all the performative aspects and the constraints deriving from the fabrication activity. An example of this changed correlation between academy and industry is the TailorCrete project developed between 2009 and 2013 in collaboration with 14 academic and industrial partners: the research project, financed by European funds, aimed to model and validate a 1:1 scale technological system for the production process of concrete buildings. The study covered the software part, for the discretization of morphologies, and the hardware part linked to the definition of the manufacturing tool and a new structural reinforcement apparatus proposed at defining an innovative type of concrete casting formwork for complex constructions (Fig. 1.8). The design solution involved the creation of a recyclable formwork, through the reuse of the wax, adaptable to different morphological configurations. The proposed constructive system introduced a digital procedure that considers the system performances, from the discretization of the morphologies to the subsequent extraction of the necessary information to construct the formworks. In addition to prototyping, this collaborative relationship has allowed to patent tools. The Mesh Mold project, launched in 2014, led to the definition of a wireframe structure able to function both as a reinforcing element of the concrete and as formwork. For this concept, it was necessary to devise and implement a special end effector capable of accurately and precisely expel the plastic polymer used for the construction. The extrusion device was prototyped and licensed, patent number WO/2015/034438, in collaboration with the Sika Technology AG industry (Fig. 1.7). The second phase of the research, Mesh Mold phase2, is currently underway and it involves the transition from a structure in plastic polymers, common ABS, to a metal structure made of a 3 mm iron net. Through the robotic manufacturing process, the metal wire is extruded and then bent to make knots and connections between different constructional elements (e.g., beam-pillar). The abovementioned second phase of this research project is preliminary to the definitive introduction of the technological system and its compliance with laws and codes related to concrete buildings.

The research projects, Mesh Mold and TailorCrete, testify how one of the aims of the studies conducted so far is the transfer of these computerized/robotized systems in traditional production and construction procedures. In this regard, a key role is played by the dialectic relationship with the industrial sector, not only for research funding

Fig. 1.7 *Mesh Mold*, Customize end effector for the anti-gravity 3D printing process. © Gramazio Kohler Research, ETH Zurich

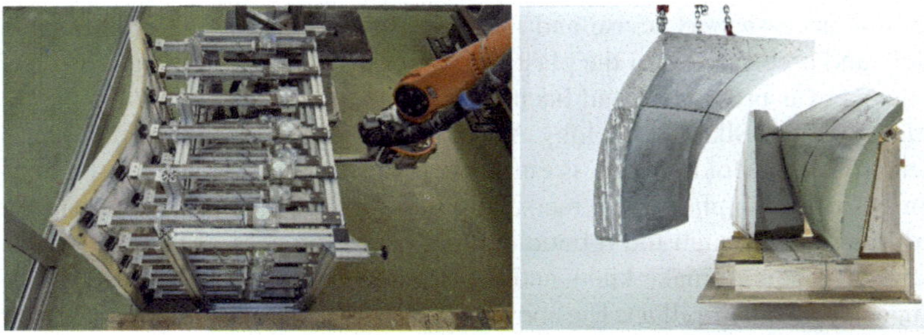

Fig. 1.8 *TailorCrete*, customized mold for free-form concrete structures. © Gramazio Kohler Research, ETH Zurich

but also for the exchange of know-how necessary to organize transdisciplinary and multi-scale testing (Fig. 1.8).

1.3.4 From Stone to Sand: The Additive Process as a Tool for Materializing Digital Complexity

A further research field at ETHZ is represented by the experiments led by Michael Hansmeyer[15] and Benjamin Dillenburger[16] on additive processes and 3D printing as tools for materializing increasingly complex morphologies thanks to the power of computational procedures for managing and analyzing large amounts of data. The algorithmic approach based on the definition of variables, constraints and associative rules, and strategies aimed at solving specific problems, is contrasted with a generative procedure that favors formal exploration without typological limitations, using the machine as a creative laboratory.[17] The focus of the experiments is on tectonic and spatial execution: the algorithm is a vehicle to explore unexpected morphological solutions able to amaze the designer himself. There is no reference to functional performance (e.g., structural) but rather to "tectonic exploration" through the definition of a calculation procedure in which the designer defines only the rules and, consequently, the composition inputs. The quantifying power allows to explore levels of resolution and topological complexity otherwise unthinkable: on the basis

[15]Michael Hansmeyer is an architect and computational designer. He is currently a visiting professor at the Academy of Fine Arts in Vienna. Previously he was Visiting Professor at the Southeast University in Nanjing, and lecturer at the CAAD Department at the Swiss Federal Institute of Technology, ETH Zurich.

[16]Benjamin Dillenburger is an architect and associate professor at the John H. Daniels Faculty of Architecture, Landscape and Design at the University of Toronto. Previously he was senior lecturer at the CAAD Department at the Swiss Federal Institute of Technology's ETH Zurich.

[17]Michael Hansmeyer, AA Video Lecture, Digital Grottesque.

of this assumption, Hansmeyer and Dillenburger define a design strategy of morphogenesis founded on simple geometric processes, traceable back to fractal geometries, able to guarantee an alternative to the standard typological classification. The geometric procedure developed by the research group, and utilized in many experiments, explores the phenomenon of surface's subdivision starting from a theoretical assumption that the same architects have coined, defined as Mesh Grammar [21]. The architects themselves explain the theory of Mesh Grammar: "The starting condition of the process is a polygonal mesh of any level of detail and complexity. The process itself consists of applying a rule to this mesh in order to produce an output mesh. This application can be repeated for a fixed number of iterations or it can continue until the mesh exhibits a specific attribute. The rule itself has two components an attribute of the mesh that is to be measured, as well as a mapping function that specifies how the measured value is converted into a transformation of the mesh. Mesh transformations consist of translation of vertices along faces' normal vectors. Optionally, transformation can include subdivision of the mesh". The process introduces a new formal vocabulary that starts from the variation of polyhedral rather than primitive geometries, whose spatial transformation through vertex modification and subdivisions of the faces always produces a new mesh. In order to realize complex topologies based on the theory of Mesh Grammar, the research team experiments a 3D printing technique that employs a standard material, sand, together with binders to produce large non-standard pieces, with a maximum resolution, in the order of 0.13 mm per print layer. The presence of special resins makes this procedure suitable for the realization of self-supporting technological units which, once assembled, do not require support structures. An example of this methodological approach is the Digital Grottesque project (Fig. 1.9) in the occasion of the ArchiLab 2013 exhibition at the FRAC Center in Orleans, the largest 3D printing work to date.

The designers describe the work with these following characteristics:

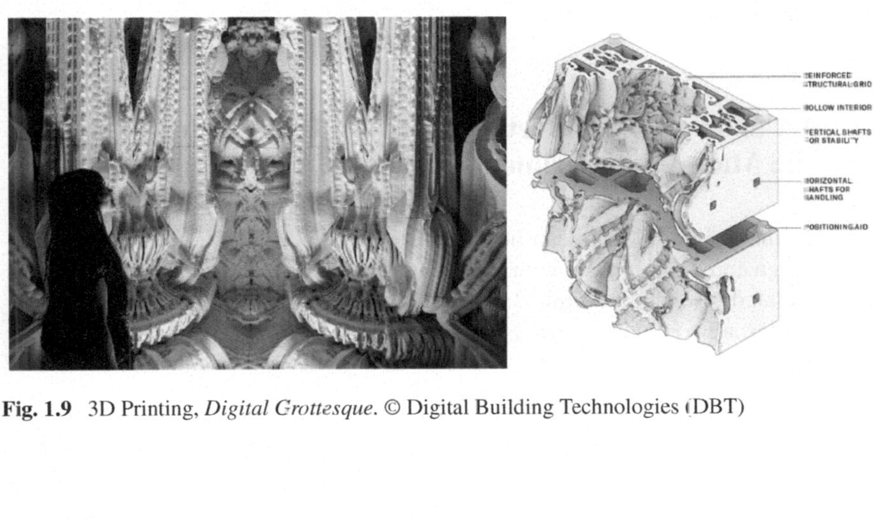

Fig. 1.9 3D Printing, *Digital Grottesque*. © Digital Building Technologies (DBT)

Virtual:

- Algorithmically generated geometry;
- 260 million surfaces;
- 30 billion voxels;
- 78 GB production data.

Physical:

- Sand-printed elements (silicate and binder);
- 16 m^2, 3.2 m high;
- 11 tons of printed sandstone;
- 0.13 mm layer resolution;
- 4.0 × 2.0 × 1.0 m maximum print space.

The description that characterizes the virtual model clarifies the following two aspects of the research:

- The Mesh Grammar approach, based on the recursive subdivision of the faces of the mesh, capable of generating 260 million surfaces;
- The computational power, able to manage and analyze an amount of data unthinkable until a few years ago.

A cave was created through the algorithm: a hybrid environment between nature and artifice, where the resolution of the details produces a visual alteration between chaos and order. A 3D printing process with sand and binders was used to materialize this topological model. This fabrication technique allowed the realization of a structure that reaches 3.2 m in height and 16 m^2 in extension, discretized following maximum dimensions of the machine, structurally performing, and with a highly detailed resolution. The proposed methodology presents an exploratory approach, where the power of computation becomes a non-standard topological generation tool, which takes shape thanks to innovative production methods that guarantees the best detail resolution.

1.4 AA School of Architecture, Between Digital Utopias and Material Experiments

Although the London School can be considered a pioneer in the field of computational design and digital manufacturing, the introduction of anthropomorphic robots and the related research line linked to their use is recent, and the experimentations on the matter are conducted mostly in partnership with start-up companies such as

Robofold[18] and *Odico*.[19] The studies carried out at the Architectural Association on the above-listed topics can be analyzed starting from two macroareas: the first one belongs to the design-oriented experiments conducted at the Bedford Square headquarters within the Design Research Laboratory (AADRL[20]) aimed at testing new material systems concerning informed and optimized arithmetic geometries. The second line of the study concerns the applications and utilization of wood as a non-engineered natural material available locally for the construction of 1:1 scale architectural structures, mainly developed at Hooke Park in Dorset.[21] The two research paths share the same tools with almost opposite ends. The first one is design-oriented, with a particular focus on complex geometric systems (e.g., developable structures) and a high level of experimentation within a techno-utopic view. The second approach aims to explore technology innovations and how they concur in expanding the possibilities of using non-engineered natural elements in the construction process. Within these two research paths, the aspects related to the material systems are investigated in terms of geometric and structural optimization and through the information of the procedure guided by the material's sustainability. The employment of industrial robots in the fabrication process enables speculation on in situ assembly logic for the realization of temporary structures aimed at increasing the public spaces of a contemporary city. The Growing System Project,[22] which demonstrates the design-oriented vocation of research, uses the industrial robot as an intelligent manufacturing agent which materializes a computational evolution of morphological generation and structural maximization. Here the constitutive and spatial performances become design inputs and play an important role in the definition of the final morphology, together with the limits and potentials of the fabrication procedures represented by 3D printing process (Fig. 1.10). In addition to the performative parameters listed, the study focuses on the material system intended as a relationship between material and structure: the chosen spatial solution and the manufacturing method adopted required the employment of plastic elements to guarantee a rapid solidification during extrusion and created a spatial formation that overcomes the logic of simple surface overlaying. Specifically,

[18] *Robofold*, was founded by Gregory Epps in 2008 following studies and experiments conducted on folding geometries as a design method. *Robofold* produces software and production technologies assigned to bend metal sheets.

[19] *Odico Formwork Robotics* was founded in April 2012 to bring robotic manufacturing outside research laboratories so it would go beyond simple prototyping. The production technology is based on the Hot Wire Cutting method and on the creation of customized software that connects the digital model and the production process.

[20] AA DRL, Design Research Laboratory, is a Master of the Architectural Association in London directed by Theodore Spyroupulos, which has always been on the cutting edge of issues related to computational design, digital fabrication, and interaction.

[21] Hooke Park is a research and experimentation centre located in the middle of a 150 ha forest in Dorset (southern part of London). For the past 30 years, the research centre of the AA School of Architecture has dealt with experimentation on wooden structures and in particular on the use of non-engineered natural material.

[22] The Growing System project is the result of research carried out within the AADRL Master (Design Research Laboratories) on the topic of robotic manufacturing and additive 3D printing processes.

Fig. 1.10 *Osteobotics*, developing a growing structure through the robotic fabrication process. © Mattia Santi

the substance used is Polycaprolactone (PCL), a biodegradable synthetic polymer which responds well to the action of solvents such as water, salt, and sodium. The computational strategy adopted aims to explore the potential of structural optimization in addition to the possibility of real-time data to adjust the variations generated by the manufacturing process during its development, compared to the digital model in terms of geometric tolerance. In conclusion, the project represents the synthesis of the theories underlying the research: development of complex and maximized structures derived from the analysis of composite, self-organized systems; adaptive production procedures based on a feedback loop logic between digital and physical model through the use of real-time data; and the study of material arrangements and non-standard building processes.

The other aspect investigated by the London School concerns natural wood linked to digital fabrication techniques and employs in the construction of 1:1 scale architecture. The natural and non-engineered material, present in great abundance in the Dorset campus forest, is used by exploiting its physical and mechanical properties without resorting to industrial processes to eliminate mechanical and structural imperfections [38]. The importance of a data-driven strategy in the design procedure is evident: in the projects of the Biomass Boiler House and the Wood Chip Barn, the scanning of the shrubs and the creation of a true catalog of the items concerning dimensional and morphological parameters represented the heart of the whole process [42]. In the Biomass Boiler House, through the 3D scanning and reconstruction of the elements, it was possible to realize a non-linear surface starting from

Fig. 1.11 *Wood Chip Barn*: use of non-engineered natural materials and robotic fabrication for the materialization of complex structures. © Zachary Mollica

the natural curvature of the timber, avoiding costly bending procedures. The computational process made it possible to integrate data regarding the material used, the structural performance, and the variables of the manufacturing site into the project, obtaining an informed morphology able to exploit the potentials of the natural, non-engineered elements based on the listed criteria. The recent realization of the Wood Chip Barn's structure began with the photogrammetric survey of 240 natural tree trunks, revisiting traditional methods of filing natural elements for making boats.[23] The subsequent scanning of 20 logs allowed to generate a database, which was employed to design the final structure utilizing the potential offered by robotic fabrication to create custom joints, difficult to achieve with standard production methods (Fig. 1.11). While sharing the same instruments, the two research paths have substantially different purposes. On the one hand, we have a design-oriented approach aimed at designing optimized and temporary architectural, on the other we have the study of a natural non-engineered material and its implementation in architectural constructions exploiting its physical and mechanical properties through a highly performing building process, and the use of data to inform the entire design supply chain.

1.5 Harvard GSD, Robotic Manufacturing for the Innovation of Traditional Industrial Production Processes

The research, conducted at Harvard University Graduate School of Design (GSD) by the Robotic Design Group (RDG) led by Professor Martin Bechthold, stands out for its pragmatic approach, which is at the base of the study and the analysis of traditional production processes for the implementation of the same, as well as for

[23]On the Encyclopédie Méthodique of 1782 there are some tables illustrating the filing of wooden elements for the construction of boats.

the examination of responsive morphologies in relation to environmental and structural performance. The focus is mainly on the innovation of industrial manufacturing procedures and traditional materials in addition to experimenting new formal qualities and enforce the efficiency of the entire assembly chain [5, 6]. The experiments involve both the software part, for what concerns the direct connection between digital process and fabrication, and the hardware part, related to the design and prototyping of end effector according to different production methods and components. Starting from the scrutiny of the failures related to the introduction of the robots in the AEC sector and the experiments on automation of the '80s in Japan, the strategy developed by the RDG opens a substantially opposite vision. The researches carried out were aimed at using an industrial machine, an anthropomorphic robot, as a necessary tool to expand the possibilities offered by traditional design and fabrication techniques [4]. The automation of processes and the mere increase in production efficiency is accompanied by an exploration of the potential offered by introducing robotic manufacturing together with a generative computational procedure that utilizes performances as project inputs, within an iteration logic and data-driven proceeding. Innovative fabrication methods have been tested in these directions, implementing the standard constructive lines. An example of this approach is the experimentation of a new type of concrete casting and forming formwork capable of exploiting the high degree of 3D space freedom offered by the industrial robot's six rotation axes on which the formwork is mounted to contain the casting.[24] This adopted design method allows to drastically reduce the cost of formwork production directed at generating highly customized and performative morphologies [1]. A further study on the subject concerned the design of a dynamic mold to be focused on realizing technological units through an adaptive procedure that facilitates the customization of the forms optimizing the material resources.

The investigations and the relationship with the industrial sector are based on a research protocol that includes the following steps [6]:

- Analysis of architecture, technological, and material system object of the study;
- Analysis of the technical aspects of the manufacturing process;
- Investigation of the production, distribution, and economic evolution;
- Summary of inquiries and formulation of the potentiality/criticality framework;
- Definition of possible design and fabrication procedure implementations;
- Composition of a material system based on the customization of the basic module and the aggregation of the constituents.

The RDG explored the research protocol described above on the production of customized ceramic systems through the collaboration with the Spanish industry "ASCER Tile of Spain". The cooperation was a result of the need to introduce a greater number of degrees of freedom in the industrial fabrication processes, ensuring

[24]The Robotic Casting construction system was tested during a workshop led by M. Bechthold, Nathan King, and Stefano Andreani during the 2012 RoboArch International Conference at the Gratz University (TU Gratz). At the end of the workshop, a horizontal element consisting of the assembly of the individual customized modules was built through the proposed production system.

Fig. 1.12 *Stone Shell*, FEA analysis, Finite Element Analysis. © Martin Bechthold

greater flexibility of the performance parameters used as a design procedure guide and, at the same time, guaranteeing the economy and sustainability of the creation process. Following this strategy, the studies were conducted with a holistic,[25] multidisciplinary approach, tackling the problem from all points of view: from life cycle analysis up to the installation and possible dismantling, considering the optimization of resources and the reduction of waste materials. As a result of this fruitful collaboration, different projects and prototypes were produced. The applied researches demonstrate how it is possible to innovate the manufacturing process introducing robotic arms as tools for shape customization and the maximization of material resource, without upsetting the entire production procedure, but working on the innovation of a specific phase inside the assembly chain instead. Like ceramics, another traditional element such as stone, has been the subject of innovation and experimentation processes. The New Stone Shells Project, developed from 2007 to 2009, demonstrated the potential of the fabrication technique sided by an informed computational procedure for the realization of performative surfaces. In the Stone Shell Project, the top layer was designed to answer two research questions. The first one was about the new expressive and tectonic qualities derived from this production method for the management of traditional elements such as stone. While the second one was aimed at identifying structural performances linked to geometric characteristics (e.g., reduced the thickness of marble slabs). The Finite Element Analysis (FEA) informed the surface through a constructional analysis which allowed to consider all the characteristics of the material (e.g., density and rigidity) and verify the structural behavior under the action of wind and weight loads [3] (Fig. 1.12). The sinuous top layer was made with marble pre-tensioned and perforated panels, fabricated through a CNC production process. The application of an end effector for WaterJet technology[26] allowed the realization of panels with different shapes and sizes. The precision of the cut and the structural analysis carried out allowed to install the

[25]The term holistic means a scientific paradigm used for the study of complex systems through an inter and multidisciplinary approach.

[26]The WaterJet cutting technology allows cutting to be achieved thanks to the pressure of 620 Mpa of a liquid composed of water and high-speed abrasive powder. The cut material is removed by abrasion and erosion.

boards without the aid of structural support. As previously stated, particular attention was paid to the software, through the design of a customized digital interface able to simplify the progress of data transfer from the virtual model to the manufacturing procedure. Another example of this approach is the research project of Andrew Payne,[27] a forerunner in studies focused on the collaboration between man-machine and the simplification of the data convey procedure for the control of the robotic arm. After reviewing the difficulties encountered in using digital programing and simulation software of the robotic manufacturing processes on the market, the researchers developed a procedure of direct iteration between the robotic arm and a scale model equipped with a series of digital sensors. The sensors were able to transfer the movements from the physical space to the electronic one and then control the robot. As a result of this methodology, an analogical robot programing approach became possible: the direct instructions on a physical model are transferred back to the machine due to a software which records the sequence of movements and then redirects them to the anthropomorphic arm employing the appropriate system of reference. This study allows the robot to be programed without specific knowledge about the algorithmic design, reducing the gap between the digital model and the physical production process.

1.6 ICD, from the Morphogenetic Process to the Machine Computation

The research conducted at the Institute of Computational Design of Stuttgart (ICD) presents an approach linked to the theories on morphogenetics [23, 32, 33] and emerging systems for a formal generation linked to innovative fabrication methods such as robotic manufacturing. Through the pavilions which summarize the studies carried out annually by the ICD in collaboration with the Institute of Building Structure and Structural Design (ITKE), it is possible to evaluate the yearly research developments. The examinations and the applied studies that followed over the years have affected the progress of material systems whose formal generation is a direct consequence of the relationship between the components and the manufacturing method, within a logic defined by Achim Menges as machine computation, constantly evolving compared to technological evolution [36]. The experiments involved the development of complex morphologies generated utilizing traditional elements, wood for example, and unconventional materials such as fibers and composites, exploiting in both cases the properties of the materials themselves [37]. The combination of the computational process and the digital fabrication becomes the means by which to investigate ways of optimally using elements commonly utilized

[27] Andrew Payne founder of LIFT Architects in 2007, is a researcher at Harvard University GSD. He developed, together with Jason Kelly Johnson, the Firefly plug-into interconnect the generative modeling software Grasshopper with the Arduino microcontroller and established a continuous interaction between the physical world and the digital environment.

in the construction sector, making the most of the physical and mechanical characteristics of the latter. A decisive role is played by the interdisciplinary approach to research, through the collaboration of other professionals for the study of biological phenomena and their transfer in the architectural field and the structural behavior and the engineering of non-linear and complex buildings as well as the development of innovative materials and production processes. The research protocol outlines a series of cornerstones. First of all, the importance of biology and the investigation of complex natural systems to extract the constructional and mechanical principles that ought to be transferred in the design procedure. Secondly, the material as an active agent is able to play an important role in the generation of form and structures through its physical traits. Last but not least, the adaptive production method able to work in accordance with the digital model and to expand the morphogenetic space of possible design solutions. Some fundamental features of biological compositions are transferred to the architectural project through the calculating process: heterogeneity, as an adaptive property of morphology relative to variable geometric parameters and external dynamic conditions; anisotropy, to define structural principles in relation to the distribution of mechanical stress; hierarchy, as a non-linear mode to aggregate components. The study of biological apparatuses and the subsequent transfer to architecture is carried out according to a bottom-up approach as explained in the article "From Nature to Fabrication: 2012 Biomimetic Design Principles for the Production of Complex Spatial Structures" [25]. As we see in this article, the understanding of biological organisms, for what concerns the biomechanical and morphological systems, is followed by the abstraction of the fundamental concepts and then continues with a technological implementation to promote an architectural redirection. As a connection between the electronic model and the physical prototype, digital and robotic fabrication are primary tools for managing the complexity deriving from the introduction of biological principles in the design procedure. Furthermore, such instruments can help in overcoming geometric and morphological restrictions imposed by the industrial assembly process leading to mass-production of the same constructive element [29]. Thanks to robotic manufacturing seen not simply as a tool to materialize a virtual model, but rather as an active agent, it is possible to expand the options offered by the morphogenetic procedure. The terms "Morphospaces of Robotic Fabrication" and "Machining Morphospaces" [33] indicate the parametric correlation between the morphological data and the robot setup. This associative relationship introduces a new space of design possibilities linked to the local scale of connection between the single elements that make up the global morphology. Another focus of the study is represented by the investigation of components based on the analysis of their physical and mechanical properties. Such assets can be seen as opportunities in the design process of informed morphologies. Taking into consideration wood, for example, the resources of the material which are usually eliminated through engineering procedures can become design inputs to define a shape. As stated by Achim Menges in the 2012 article "Material Resource Fullness": "In addition to the ecological incentive, the intricate make-up and complex behaviour of wood provides another reason for employing computation to explore its materially inherent design opportunities", digital computation becomes the mean by which to discover the material's possibilities in relation to its composite and non-linear conduct. To this extent, features such as elasticity, anisotropy, and irregularity become guiding principles of the computational process, and through simulation tools, it is possible to integrate them throughout the entire design chain. For the research pavilion built in 2010 within the performative wood research line, the physical properties of the material were incorporated into the calculating procedure through a series of tests carried out on the elastic behavior of the plywood (wooden multilayer panels). The structural features were extracted in order to be integrated into the computational process for the simulation of the morphology and the consequent constructional examination. The toroid generated by the simulation of the individual wooden panel's elastic properties was fabricated by combining the manufacturing and assembly characteristics into the calculation procedure (Fig. 1.13) with 500 geometrically different parts. The aggregation of the stripes and the structural joints control the shear-resistant connections as well as the linkages between the building and the base, which acts as a constructional foundation. The final pavilion was made entirely with 6.5 mm thick wooden stripes whose elastic features were started up on-site through the application of a mechanical force, generating a light and structurally efficient morphology. The internal and external perimeter was defined by a wooden construction for the structural anchoring of the strips.

Another line of research concerns the investigation of how the biological principles of fibrous systems, defined as Fibrous Tectonics [35], can be transferred into architecture: "Fibrous systems are omnipresent in biological structures, but the fibers are made from only a very small number of fundamental materials... From this limited palette of fundamental constituents, nature manages to build an incredibly diverse range of systems by varying the fibrous arrangement and density, as well as the chemical composition of the matrix". With these words in the 2016 article on Fibrous Tectonics, Achim Menges and Jan Knippers describe the characteristics of fibrous systems indicating their performance and responsive ability, as a potential to explore and redirected into architecture. Starting from these properties, research projects have been developed to explore the structural and formal potential offered

Fig. 1.13 *ICD Research Pavilion 2010*: form-finding process and on-site installation. © ICD/ITKE University of Stuttgart

by the composite materials complex in terms of computational process and new production methods.

1.6.1 Research Pavilion 2013/2014: Fibrous Tectonics, Exploring New Innovative Materials and Production Technologies

The 2013/2014 research pavilion explores biological principles [25] for the definition of an innovative manufacturing methodology for multilayer technological units realized through the winding or overlapping, of fibers able to minimize the need for supports during fabrication and to guarantee, in any case, high customization of the shape. The constructional concept derives from the abstraction and the subsequent conveyance of biological principles observed in *Elytron*, a protective shell on the wings of cockroaches, classified by the presence of a multilayer structure and the anisotropy of the material that has different levels of chitin64 depending on the different distribution of mechanical stresses. The morphological responsiveness, in the case of *Elytron*, is determined through the diversification of the component properties and the structural hierarchy by the presence of the two different layers that make up the shell of the animals. In particular, the constructional performances of this biological structure are guaranteed by the mechanical features of natural fibers as well as by its morphological conformation organized on two superimposed planes. These parameters were transferred to the project to construct structural modules made up of natural fibers, using constructional resins to differentiate their properties (Fig. 1.14).

The manufacturing process employed two collaborating robots (Fig. 1.15) which had as end effector the metal settings used to weave the differentiated filaments according to their function. Glass fibers were the support edifice for structural carbon

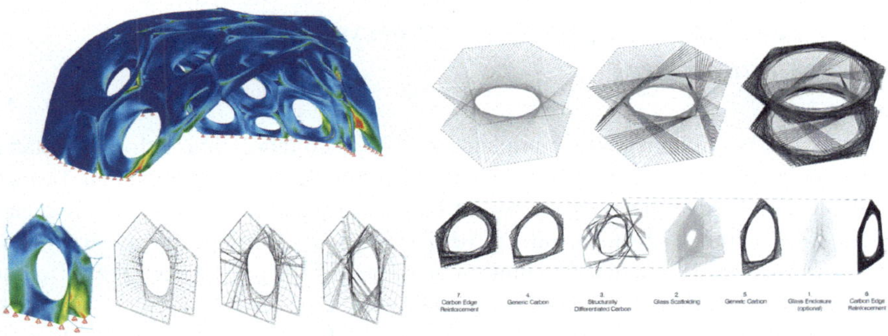

Fig. 1.14 *ICD Research Pavilion 2013/2014*: distribution of the material following the structural analysis and diversification of the natural fibers with regard to their mechanical properties. © ICD/ITKE University of Stuttgart

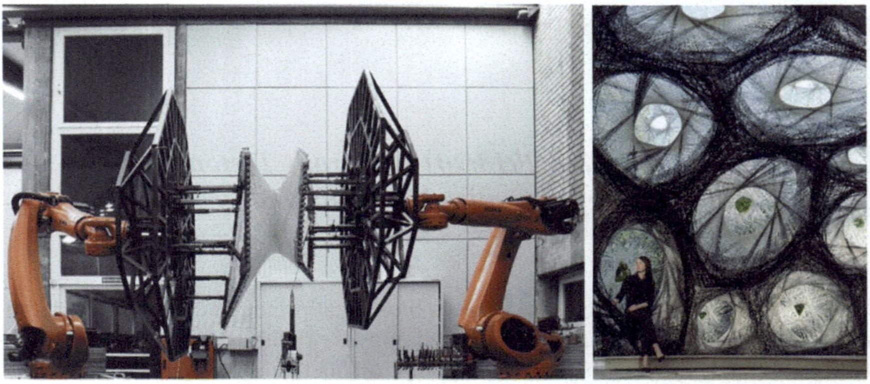

Fig. 1.15 *ICD Research Pavilion 2013/2014:* manufacturing process through the articulated movement of two robotic arms used for winding the fibers. © ICD/ITKE University of Stuttgart

fibers arranged on the edge of the frame and along the load distribution lines. Each hexagonal component was the result of adaptation to specific load conditions that determined the fiber's different spatial placement and density. The parameters of the manufacturing process, related to the material system, were included in the computational process together with the simulation of the structural analysis; the latter played a fundamental role. The largest constructional module was 2.6 m in diameter with a weight of only 24 kg, demonstrating the achievement of a light and performing complex. Through the assembly of 36 modules, a 50 m^2 pavilion was produced with a total weight of 593 kg, reifying the efficiency of the structural and material system made with industrial robots.

1.6.2 Research Pavilion 2015/2016: A Manifesto for an Informed Architecture Between Biomimetic Principles, Morphogenetic Process, and Robotic Manufacture

The research pavilion, recently completed by ICD in collaboration with ITKE, starts from the examination of discretized shell structures defined as Segmented Timber Shells and is characterized by the use of laminated wooden panels with a reduced thickness that ranges from 3 mm up to 6 mm. Given the type of constructions, the studies involved a detailed definition of a class of hinges between the different modular elements that make up the morphology at the architectural scale. The joints, which in traditional wooden structures are optimized for units of consistent thickness, must be reinterpreted to be in line with the specific structure and the reduced size modular elements. The need to maximize the connections between the thin structural segments, represented by 6 mm minimum thickness, and optimal distribution of the

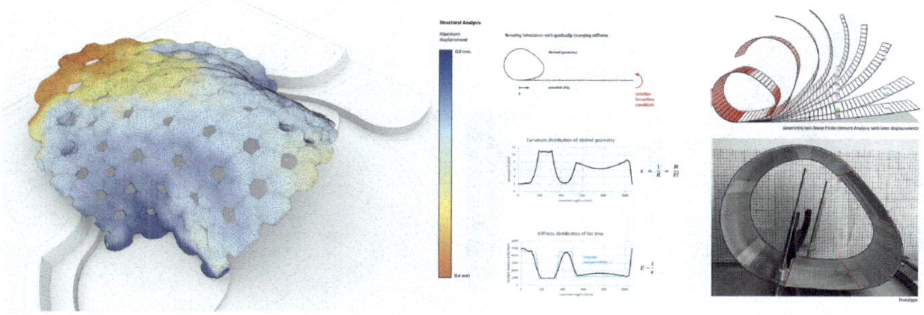

Fig. 1.16 *ICD Research pavilion 2015/2016*: structural analysis and bending process of thin wood panels for the construction of individual technological units. © ICD/ITKE University of Stuttgart

material, required a transdisciplinary design approach based on the observation of non-standard processing implemented through robotic manufacturing tools (which guarantee adaptability of the system). The research involved the study of biological systems, specifically the *Echinoidea*,[28] to understand and analyze the structural and morphological organization depending on different scales, and define:

- Type of hierarchical distribution, tessellation, of the elements that discretize the global morphology;
- Type of connection between the components on the global scale;
- Definition of multi-material connections;
- Distribution of the resistant section of materials related to the diffusion of structural forces.

The models conveyed in the project through the studies conducted on the *Echinoidea* were defined following the above-listed categories (Fig. 1.16).

In particular:

- Transfer of the natural growth mechanism of the *Echinoidea* through a form-finding process based on the phenomenon of the circle packing to reach a high density of connection between the components and to guarantee structural rigidity;
- Definition of volumetric constructional prototypes, with a diameter ranging from 0.5 to 1.5 m, made with laminated panels with reduced thickness, from 3 to 5 mm;
- Double level of association between the structural elements: finger joint and elastic connection guaranteed by the seam between the various components, made utilizing a fabric thread and a collaborative procedure between the robotic arm and a traditional sewing machine (Fig. 1.17);
- Angular variation of laminated wood fibers depending on the distribution of mechanical stress.

[28]The Echinoidea, belongs to the class of Phylum Echinodermata which includes marine organisms defined as sea urchins.

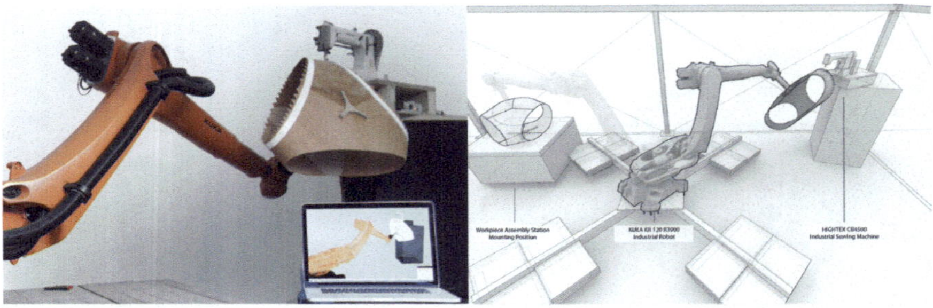

Fig. 1.17 *Cyber-Physical Making*: collaboration between the industrial robot and a traditional sewing machine. © ICD/ITKE University of Stuttgart

Once the morphological parameters to be transferred to the project were delineated, a production technique for the fabrication of the double-layer laminated wood panels (previously formed through bending) was developed. The cooperation between the robotic arm and the textile machine was necessary to transmit the sewing techniques typical of the textile procedures in the architectural project.

The manufacturing process was organized based on the following steps:

- Lamination of wood;
- Subtractive procedure, CNC milling, to define the shape of the components and to realize the first type of connection represented by the finger joint;
- Bending process of single laminated sheets and temporary fixing of the same on the robotic arm in anticipation of the sewing process;
- Sewing procedure of the individual components using elastic laces for the connection of the individual technological units on the installation site;
- Final on-site assembly (Fig. 1.18).

In addition to the customization of the components produced, the innovation introduced by the manufacturing process is represented by the possibility of combining typical textile processing techniques with standard manufacturing procedures, thanks

Fig. 1.18 On-site assembly process. © ICD/ITKE University of Stuttgart

to the adaptability guaranteed by the industrial robotic arms. The option of sewing, and subsequently connecting the technological components utilizing threads, laminated and folded panels, ensures structural stiffness as well as avoids the criticalities of the traditional component's bonding process (such as the delamination of the panels and the instability of joints, due to their curvature before assembly). At the end of the fabrication procedure, 151 panels with a thickness that ranged from 3 mm up to 6 mm were produced. By the end of their assembly process, a spatial structure of 85 m^2 with a structural span of 9.3 m and a total weight of 780 kg was obtained. The technical data described above and the manufacturing and installation times show how the application of the proposed methodology represents a valid alternative to the standard industrial construction procedures for the realization of light and structurally efficient wooden compositions. Furthermore, an element like wood, traditionally used in the construction sectors, is revisited through inventive production methods, robotic manufacturing, and a cyber-physical making process. The above consideration opens experimental scenarios for the application of innovative formal codes related to performative parameters not yet explored in the logic of optimization of material resources.

1.7 MIT, from the Synthetic Assembly Process to the Concept of Biological Growth

The studies carried out at the Massachusetts Institute of Technology (MIT), specifically by the Mediated Matter research group led by Professor Neri Oxman, demonstrate the peculiarity of a bidirectional relationship between biological sciences and computational design, as well as the digital manufacturing. The experiments performed by the research group in the last few years, begun with the analysis of the design and production process rooted in the artificial assemblage of parts in comparison with the natural world linked to growth logics and self-organization principles. The theory compares a synthetic approach, built to be introduced in the natural environment, and a biological approach, where nature itself creates by following an evolutionary logic: from the assembly of parts of modern derivation to the principles of growth based on Darwin's evolutionary theory. This methodological method is based on multidisciplinary collaboration between different skills and combines four components: computational design, additive process, component's engineering, and synthetic biology. The proposed approach aims to combine a top-down with a bottom-up method, derived from digital calculation techniques and biology and natural system's theories [48]. The whole experimental procedure presents a new awareness concerning materiality and materialization in the creation of artifacts, starting from the anisotropy of elements present in biological organisms, which allows for the diversification of functions based on the variation of material's properties, such as density and elasticity. The study of the material and its variation of the traits depending on performative criteria represents a new level of material perception. This awareness begins with the observation of the natural world where the formal generation through a form-finding process is guided by the intent to achieve maximum performances with the minimum use of components resources [45]. Through digital computation tools, it is possible to program the material and transfer the anisotropy properties of biological elements to artificial products which can be materialized through digital fabrication methods, particularly the additive process of 3D printing. The possibility of programing and manufacturing materials whose functional features can be specified at the local scale gives rise to distinct research trajectories, which correspond to types of anisotropy found in nature such as

- Geometric anisotropy determined by density and spatial organization;
- Structural anisotropy settled by the distribution and orientation of the resistant elements;
- Material anisotropy assured by the heterogeneity of components.

From these theoretical and methodological assumptions, and with the help of an interdisciplinary project team, the investigations conducted involved the development of software, the prototyping of tools, and end effectors necessary to fabricate the components up to the design of material and construction systems. Through the algorithmic software, it is possible to program the material and insert the performances as design inputs are able to determine the mechanical and physical characteristics of the components (which can later be produced with the advances of digital fabrication). What is defined as "Functionally Graded Digital Fabrication" [45] mainly refers to the fabrication procedure of elements whose structural, physical, and mechanical properties may vary according to their function. These synthetic naturally inspired materials cannot be produced with the traditional CAD-CAM design and manufacturing technologies, as they need customized tools and experimental production processes able to guarantee their heterogeneity and functional differentiation. An example of this investigation path is the experimentation done for the design and prototyping of an end effector to be applied on a robotic arm capable to extrude anisotropic concrete. In this concrete, the heterogeneity expressed by porosity is programed in a virtual environment to obtain a light and structurally performative structure (performative in terms of optimization of the distribution of mechanical stress within the structural component). The heterogeneity of the material can be found in nature, especially in bone tissues where the variation of the distribution and the components density occurs following the main lines of mechanical stress. Another strategy to be applied to informed engineering elements is represented by the variation of chemical compounds, like collagen and cytosine, in soft collagenous tissues, which corresponds to the modification of mechanical stress in the bone. The biological principle is transferred into the architectural project utilizing an instrument which allows mixing polymers. When such polymers are combined in different percentages, they create distinct mechanical properties. This kind of tool applied on a robotic arm allows the realization, through an additive process, of monolithic structures with different mechanical assets; such complexes can be programed and customized through digital computation [44].

1.7.1 Robotic Additive Manufacturing of Lightweight, Biodegradable and Materially Heterogeneous Structures Across Scales[29]

The project can be considered as a manifesto of the Mediated Matter theories on the relationship between biology, digital computation, additive manufacturing, and materials engineering. The calculating procedure integrates the virtual and the physical fabrication process into a single virtual platform [48]. The experimentation involves basic biopolymer such as chitosan,[30] a polysaccharide composed of D-glucosamine and N-acetyl-D-glucosamine joined by β (1–4) bonds, traceable in abundance in nature, together with chemical agents and water. The solution, at the base of the experiment, is comprised of chitosan, in a quantity varying from 3 to 10%, with 1% of aqueous solution of acetic acid (CH3COOH), sodium alginate (C6H9NaO7), 5% cellulose (C6H10NaO5) (compared to the solution with 10% chitosan), and 16% chitin powder (C8H13O5) n [41]. The combination of the elements listed above produces a variety of fluid-dense materials characterized by different physical and mechanical properties such as structural rigidity and transparency level. The water evaporation process guides the final morphology of the prototype while the presence of a bacterium programed in the laboratory enables converting carbon dioxide (CO2) into oxygen (O2), generating a real process of biosynthesis able to increase the structural mesh. The inquiry that led to the creation of the prototype involved several levels, starting from the software platform that allows managing, through a script,[31] data related to the positioning, the synchronization of the material extrusion, as well as the component's metadata. While the manufacturing procedure involves the setup of a robotic processing cell and the design of a custom-made end effector for the extrusion of fluid-dense biopolymers composed of pneumatic elements, and multiple nozzles[32] capable of conducting the deposition of the material depending on its physical and mechanical properties and following the digital model [39]. Thanks to the direct connection between the digital model and its manufacture, it was possible to place the material based on structural-mechanical and optical traits, generating a 12 m tall prototype characterized by lightness and structural rigidity. After the production, the prototype was subject to a drying process whereby the evaporation of the water guided the generative and formal procedure. The material used, composed of a biopolymer, water, and chemical agents, is completely biodegradable. Water is a form-generating element through the evaporation process, and at the same time, it is the element that allows the polymer to dissolve again in aqueous solution.

---

[29]http://matter.media.mit.edu/environments/details/water-based-digital-fabrication-platform.

[30]Chitosan is formed by treating chitin, generally obtained from the exoskeleton of crustaceans (crabs, shrimps, etc.) with a basic aqueous solution.

[31]Specifically, the code that controls the geometry and the manufacturing process was realized in the 3D Rhinoceros modeling environment and C++ programming language.

[32]The term nozzle refers to the nozzle that manages the extrusion of the material in 3D printing processes.

1.7.2 Gemini[33]: *Digital Computation and Additive Manufacturing for a New Relationship Between Architecture and Biology*

The Gemini chaise longue[34] [47]is the manifesto of the research line proposed by Professor Neri Oxman, invested in designing by following biological principles while taking advantage of innovative digital manufacturing tools. Starting from the study of the Ornithogalum plant,[35] better known as the Star of Bethlehem, and its cellular structure, which regulates and optimizes its surface-volume ratio in relation to execution variables, the chaise longue design is aimed at the achievement of acoustic performances through the differentiation of composition and material traits [48]. The designed seat presents performance parameters at different scales. At the scale of the object, the complex morphology was conceived as a semi-closed aniconic chamber that promotes the propagation of sound. At the detail scale, which corresponds to that of sound waves, the external surface was conceived as a double curvature cell system, which is interconnected to favor the absorption of sound based on their surface-volume ratio. At the material composition level, the differentiation of the properties was managed by a computational process [48]. The distribution of material on the seat's external surface is informed by a series of parameters related to the distribution of a single person's weight, to optimize comfort and the absorption of sound from external sources. Lastly, the production procedure is the result of the combination of two fabrication technologies: subtractive process, milling, for the semi-closed wooden structure, and an additive procedure, multi-material 3D printing, for the construction of the external surface. The variable traits of the material, together with the curved geometry of the individual cells, have allowed managing the properties of sound absorption locally.

1.8 TU Delft, Robotic Fabrication, and 3D Printing for Informed Architectures

The Technical University of Delft is one of the most important universities that is actively working on the topic of informed architecture. Their research can be divided into two main paths: the first one concerns the use of 3D printing for the fabrication of informed and optimized technological units according to environmental and structural parameters; while the second investigates the possibilities for the exploration of informed tectonics offered by the introduction of robotic manufacture.

[33]https://www.media.mit.edu/projects/gemini/overview/https://www.media.mit.edu/projects/gemini/overview/.

[34]The Gemini chaise longue was created by Professor Neri Oxman in collaboration with Prof. W. Craig Carter of the Department of Materials Science at MIT in Boston.

[35]Ornithogalum is a genus of plant of the Liliaceae family widespread in various parts of southern Europe, Asia, and South Africa.

The studies of the Design Informatics Department are focused on the maximization of energy and environmental performance, specifically in terms of acoustics and daylighting, through design and fabrication of informed technological systems. Through the employment of computational tools it is possible to link in the same workflow the optimization of performative benchmarks, according to geometric variables and constraints defined by designers, and the fabrication process, taking into account the parameters which regulate production as well as employed materials. The experiments are focused on the exploration of new tectonic possibilities introduced by the additive procedure, and how the latter can be utilized to maximize execution and design morphologically informed systems. Through this methodological approach, the final form turns out to be the result of a consequential process able to absorb and process data related to the geometry, the material, and the production procedure. Among the research objectives, one of the main ones is the attempt to start a delocalized fabrication process where everyone can produce his technological system by assembling the basic units, which can be printed using a common desktop 3D printer. Through digital platforms, the user can customize his/her technological system depending on a specific application context, while respecting a series of variants established through a designer-defined space of design possibilities. Another research path is represented by robotic fabrication[36] applied to architecture conducted by the Hyperbody Research Group led by the Project Leader Henriette Bier. The focus of the experiments is the integration of robotic production throughout the entire design and construction process, from modeling to simulation to fabrication [7]. Such integration is aimed at creating visually static informed architectures. In this type of proceeding, performance becomes the guiding parameter of the generative procedure, and the resulting tectonic complexity can be materialized through digital manufacturing technologies, such as robotic fabrication, which guarantee greater flexibility in production processes. By taking advantage of the technical characteristics of the tool utilized, various manufacturing methods were investigated, ranging from the 3D printing additive process, using plastic or clay polymers, to subtractive procedures, specifically milling and hot wire cutting. Through an integrated process defined as design-to-production, it is possible to realize structures with the following characteristics:

- Customized and diversified according to material and fabrication procedure;
- Energetically efficient;
- Structurally performative;
- Morphologically responsive.

The production of informed architectures as established above needs the realization of scale 1:1 tests with a flexible virtual platform able to manage the complexities derived from the computational process while guaranteeing manufacturing flexibility in terms of processing methods and tools and materials used. In order to explore the

[36]Robotic Building is a three years research program of the Hyperbody department at the Technical University of Delft in collaboration with 3TU Bouw, ABB Robotic, KUKA, 100% TUD, Delft Robotics Institute.

possibilities offered by this innovative design approach for engineering and fabricating hybrid performance surfaces according to different scales, materials, and technological systems, a series of topics have been inquired. The researches involve three scales of detail: macroscale, to define the topology of the structure; mesoscale, suitable for discretizing the global shape and determining the components; and micro scale, whereby the structure is settled according to variable criteria, such as porosity. In general, the research aspires to define the following relationships:

- Macroscale—structure—topology;
- Mesoscale—components—assembly;
- Microscale—environment—porosity.

The results of the design process are represented by informed technological systems capable of adapting through morphologic configuration to structural and environmental inputs and guaranteeing feasibility by considering all the parameters of the fabrication and assembly procedures.

1.9 UCL Bartlett, Supercomputing and 3D Printing

The Research conducted at the UCL Bartlett University of Architecture in London, within the MA in Architecture, MArch Architectural Design (AD), can be described by analyzing four keywords: supercomputing, informed process, discretization, and additive process. In one of the recent exhibitions organized by the London school,[37] aimed at presenting the experiments conducted during the academic year of the Master, Cluster 4[38] introduced the prototypes with the following definition: "Architecture has never been digital: despite the use of computers to calculate huge amounts of complexity, the way that we build is still analogue, and therefore our increasing computational power is merely used in representational way". The term "digital fabrication is misleading as well; 3D printing is an analogue process, similar to the way the CNC mill automates an artisanal action ". The main focus of the experiments is placed on the digital process as a virtual tool able to interconnect the electronic space with the physical one. If the digital avant-garde of the '90s produced notable advances in terms of tectonics and morphology through the use of Spline curves and Blob Surfaces [31], the same cannot be said for the fabrication procedures and their materiality. During the first digital age, the diffusion of digital simulation and animation software favored the development of new surface topologies that did not correspond to an equally profitable materialization process. In this regard, the research intends to bridge the gap between virtual and physical space through an operative methodology directed at introducing a renewed material sensitivity in electronic procedures.

[37]The B-Pro Show 2016, September 27, 2016–October 2, 2016, is an exhibition of the MArch AD (Architectural Design) and MArch UD (Urban Design) works at UCL Bartlett in London.

[38]Cluster 4 of the Master in Architecture deals with Computational Mereology—Large Scale Discrete Fabrication and is directed by Gilles Retsin and Manuel Jimenez Garcia.

Thanks to the calculating power, defined as supercomputing, it is possible to act on a digital and material model through the following strategies:

- Simulating the action of dynamic forces on a material system;
- Including a large amount of data, data-driven strategy, from different fields of investigation;
- Exploring complex networks through multi-agent systems (MAS),[39] studying natural phenomena such as self-organization principles;
- Incorporating in the computational process the manufacture and components parameters.

The research is described as follows: "Currently, one of Wonderlab's main research trajectories is engagement with GPU-run supercomputing, in which large quantities of data allow us to traverse scales and disciplines, to truly go from micro to macro, dive into new discoveries coming from material science and test them in design applications at scale. By encoding matter with algorithmic parameters—now widely practiced in the sciences and many industries such as automotive—we are working with what could be called materialisation prior to materialisation, designing not only form but possible material states before they get materialised"[40] [54]. This methodological approach intends to open the debate on the post-digital era, to experiment new tectonics, transforming the properties of materials into facts, bits, which are included in the computational model through data-driven and supercomputing processes. The resulting tectonics define an organic and fluid space, able to activate a new ecology through the study of complex natural structures, which are opposed to synthetic logics of mechanical assembly of parts. Another approach on the same topic foresees the exploration of computational design potentialities in the field of discretization systems, where the comparison between parts and whole, and vice versa, acquires value through the information of procedures and digital matter. The study on system's mereology, the investigation of properties and formal relationships between an integer and its pieces, is investigated in terms of computational power, new materiality, and a renewed sense of aesthetics, following the discretization and assembly of parts of the modern era [50]. The discretization of the surfaces is analyzed to examine the possibility of informing the material for each element, including in the calculation procedure the traits and characteristics of the material and the manufacturing process. Both theoretical viewpoints, which contrast continuity through the discretization of elements, exploit the potential offered by digital fabrication devices, specifically the possibility offered by the addition of a 3D printing procedure and by robotic construction assembly. If standard 3D printing can be considered as an analog process consisting in linear material incorporation, the combination of additive procedures with anthropomorphic robotic arms allows exploration of innovative aspects of the relationship between material, production process, and work-production space

[39] A multi-agent system consists of multiple agents present in a specific environment and interacting with each other through an appropriate organization, managed according to a set of predefined rules.

[40] http://www.w-o-n-d-e-r-l-a-b.com/vision/.

[15]. In fact, the analysis of organic structures, fluid and continuous bodies, experiments with supplementing procedures in which the logic of linear, layer-by-layer deposition is overcome by the spatial extrusion of the component, which works on the three dimensions, making use of the material's qualities and structural optimization. By taking advantage of the potential offered by robots, specifically thanks to the six degrees of freedom in the movement of the anthropomorphic arm, a vertical printing process can be defined. Such a procedure can deposit the material outside the x, y plane and follow 3D paths. Unlike 3D printing, in formal discretization processes, assemblage logics are prevalent and can be supported by robotic fabrication due to its ability to position elements in the workspace with precision in compound assembly procedures. As a consequence, digital computation and flexibility of the models turn out to be performative design tools for managing the complexity generated by the component's aggregation and, as a result of the variation of the data structure, quickly exploring different design possibilities.

1.9.1 CurveVoxel[41]: Exploration of Continuous Tectonics Through Additive Processes of Robotic 3D Printing

The *CurveVoxel* project, developed at UCL Bartlett, investigates the potential deriving from the application of innovative computational and manufacturing techniques for the production of informed and optimized spatial buildings. The theme of experimentation concerned the reinterpretation of a piece of furniture, cantilever chair or S-Shaped Chair, through the determination of a cellular structure informed and materialized through an anti-gravity and free-form 3D printing process. The computing power allowed to obtain a level of detail that goes beyond the scale of the technological unit, reaching the pixels[42] and consequently the voxel.[43] Through the definition of geometric rules, the computational process can alter the resolution and geometry of the curve due to the mechanical, structural stress in such a way as to vary the density of the 3D mesh based on structural data [15]. In order to realize this adaptive composition, the code works on the determination of tangents and the control points of the generated curves starting from the voxels which, in turn, represent the first discretization of the surface. The digital process informed by the properties of the material and the construction proceedings allows the creation of different tectonic solutions, as the inputs and the performances that influence the system to differ. In

---

[41] http://www.curvoxels.com/about. Design: CurVoxels Team (Hyunchul Kwon, Amreen Kaleel, Xiaolin Li) Tutor: Manuel Jiménez García, Gilles Retsin, Vicente Soler Senent, Research Cluster 4 at the UCL the Bartlett School of Architecture.

[42] A pixel, in computer graphics, indicates each of the point elements that make up the representation of a digital raster image, for example on a display device or in the memory of a computer. To put it plainly, the pixel is the smallest element that constitutes an image.

[43] A voxel (volumetric pixel or more precisely, volumetric picture element) is a volume element that represents a signal or color intensity value in a 3D space, similar to the pixel representing a 2D image. Voxels are often used as a basic element for the visualization and analysis of scientific data.

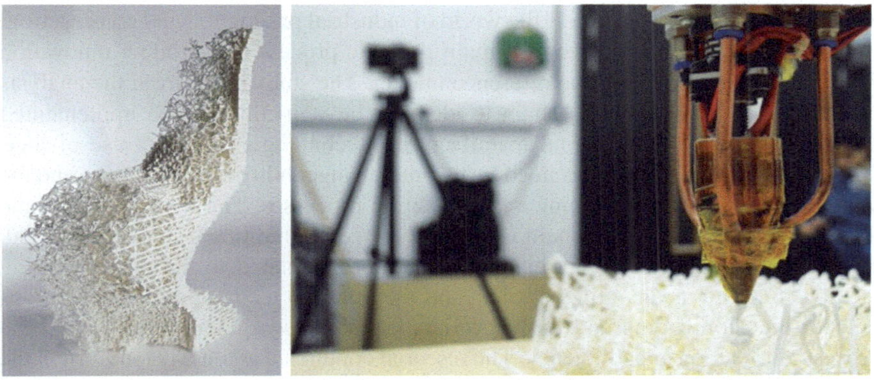

Fig. 1.19 *Curve Voxel Chairs*. © CurVoxels

order to materialize the virtual space, it was required to design and prototype an end effector for the robotic arm. As a result of the latter, the plastic polymer extrudes spatially, and not exclusively on the x, y plane. The additive process described above permitted the continuous extrusion of the morphology thanks to the development of a network of relationships between tools, material, and variable parameters, such as temperature and extrusion speed (Fig. 1.19).

1.10 Comparing Theories and Methodologies: Materials and Fabrication Techniques for Informed Architectures

The analysis presented so far showed the existence of different research itineraries and various international universities working on issues related to computational process and advanced manufacturing procedures for producing informed architectures. Given the complexity of the topic, it was necessary to identify the major study paths to outline the main theoretical and methodological approaches. Through the survey, it was possible to make a critical comparison and to highlight the differences among the research paths in terms of digital processes, production methods, and materials used. While the common thread, which links the various projects investigated, appears to be the design and fabrication of informed architectures, whose formal generation is guided by performance parameters. All the experiments showed a renewed material sensitivity, compared to the processes of the first digital age, through a continuous workflow between the 3D model and manufacturing procedure. The relationship between the computational process and innovative manufacturing methods becomes a formal exploration tool guided by the realization of certain achievements considered as design inputs through which to originate informed and optimized architectures. The choice of components and fabrication procedures is closely connected to the correlation that links research to the world of industrial production and construction: some of the experiments have as a main goal the process and product innovation for a future introduction in the reference market. The study starts from the analysis of the industrial construction chain to identify the steps that can be implemented through integration with digital manufacturing technologies, making use of their flexibility, the versatility guaranteed by the interchangeability of the utensils, and the simplification of the programing and simulation processes. The introduction of the methodology in the design chain especially concerns traditional materials (wood, ceramics, concrete) whose formal variety is closely related to the manufacturing methods and the means utilized to expand the possible technological applications related to customization and formal variation, guaranteeing and at the same time process automation and efficiency. The research conducted at the ETH in Zurich, the Robotic Design Group of Harvard University GSD, in collaboration with industrial partners, present tangible results in terms of technology transfer and management of complex processes underlining the application of technological systems in wood, ceramic, and cement in the construction sector. The modernization of production procedures also involves non-engineered natural elements such as the wooden logs used in the AA testing for the Wood Chip Burn and Boiler House, for the construction of 1:1 scale architecture. The complexity deriving from their use in architectural processes, resulting from the irregularities and properties of the wooden logs, is managed through the information of the procedures, while the production is guaranteed by the versatility and efficiency of the end effector. The link between research and industry is accompanied by an examination of compositional aspects, aimed at the geometrical-formal exploration and the study of innovative structural typologies related to the theories on tectonics deriving from the application of emerging technologies in architecture. Another aspect that should be underlined is the relationship between biology and architecture, which is transmitted to the design process through the abstraction of biological principles and their technological implementation in the latter. Studies conducted by the Stuttgart ICD, and the MIT's Mediated Matter research group, show how structural and mechanical fundamentals pertaining to biological organisms can be transferred to architecture by understanding the logic and correlations that exist between the elements, without operating mere mimesis of the form. At present, the limits highlighted are represented by the scalability of the processes and their application in compound architectural organisms.

1.11 Conclusion

The chapter establishes the theoretical framework of the "informed architectures" topic, underlining the different approaches through the definition of project variables (e.g., materials, tools) and invariants, which can be found in all the research paths. Starting from the survey, it is possible to build up a narrative founded on a series of case studies, which were selected based on the theoretical framework described above and intertwining computational strategy and manufacturing logics.

In all the analyzed projects, performative parameters constitute the design guide and the project's inputs, defining a new concept of responsiveness and adaptation dependent on information and subsequent materialization of the morphology through modern fabrication processes, such as robotics and 3D printing. The introduction of robots and the final affirmation of 3D printing technologies marked the beginning of the post-digital era in architecture, which is rooted in the combination of digital computing and manufacturing. In the post-digital era, aspects related to environmental sustainability and creativity become the core of the project, standing before automation and the economy of processes. The design possibilities offered by a direct connection between generative proceedings and manufacturing do not imply a formal evolution without logical relationships; they rather identify the machine as an agent that materializes optimized and informed geometries. Due to digital computing, performative parameters can be used as design inputs in a synoptic view, enabling the achievement of informed geometric configurations depending on the criteria used as incentive. Considering all the above, digital manufacturing technologies and materials are not variables to be considered at the end of the process, but rather conditions for the outline of the meta-project.

1.11.1 Introducing the Analysis of Case Studies

The theme of informed architectures opens up innovative scenarios related to performance-based parametric processes and data-driven strategies such as the utilization of new materials, the exploration of new production methods for traditional components, as well as the introduction of new formal codes. Considering the above-listed issues, case studies were selected verifying their compliance with the following criteria:

- Presence of a performance-based computational process through which performative parameters inform the morphology;
- Materialization of the digital model through modern digital manufacturing technologies, specifically robotics and 3D printing;
- Exploration of informed architectures, with innovative formal codes.

To facilitate the critical synthesis, we proceeded through a systemic approach where, by a system we mean the set of relations between constituents that form the whole, referring to the normative UNI 8290 classification and articulation of technological units and technical elements and the UNI 7867 standard that defines the building and environmental systems. The selected projects were analyzed according to the variables previously listed through the identification of three categories of reference from which a dialectical comparison aimed at identifying the potentials and the critical aspects of a given process could be initiated. The three reference categories were delineated as follows:

- Technological unit, describing an element that associates itself with a grouping of functions focused on guaranteeing environmental performances;
- Technological system, describing an organized set of technological units;
- Architectural organism, describing a unitary and structured set of technological apparatuses.

Once the reference macro-categories were established, the following application fields were identified:

- Internal partition;
- External partition;
- Vertical and horizontal closures;
- Bearing complex.

The systemic reading of the case studies was necessary in order to carry out a comparative analysis aimed at delineating the pros and cons deriving from the possible application on a 1:1 scale. Each case study was reviewed through the definition of parameters related to the computational and manufacturing process. The focus was placed on the computational paradigm, which determines the generation and discretization of the form, identifying the variable and constraint criteria which govern the space of the design possibilities. As for the creation of form, also for the production procedure, the main factors regulating the digital fabrication methodologies were outlined. The case studies analysis was preliminary to the critical reading which was concerned with the following points:

- How performances can be included in the early stage phase of the design of technological units, technological systems, and architectural organisms taking into account the production methodologies used;
- What are the potentials and the criticalities deriving from the introduction of the proposed planning methodology and the new production methods in architectural practice and in industrial processes;
- What are the advantages in terms of optimizing performance and material resources, with special attention toward environmental sustainability;
- How many and what are the new formal codes that pertain to informed architectures.

ICD/ITKE University of Stuttgart projects credits ICD/ITKE Research Pavilion 2013-14: ICD Institute for Computational Design—Prof. Achim Menges; ITKE Institute of Building Structures and Structural Design—Prof. Jan Knippers; Scientific Development: Moritz Dörstelmann, Vassilios Kirtzakis, Stefana Parascho, Marshall Prado, Tobias Schwinn; Concept Development: Leyla Yunis, Ondrej Kyjánek; System Development, Fabrication & Construction: Desislava Angelova, Hans-Christian Bäcker, Maximilian Fichter, Eugen Grass, Michael Herrick, Nam Hoang, Alejandro Jaramillo, Norbert Jundt, Taichi Kuma, Ondrej Kyjánek, Sophia Leistner, Luca Menghini, Claire Milnes, Martin Nautrup, Gergana Rusenova, Petar Trassiev, Sascha Vallon, Shiyu Wie and Leyla Yunis, Hassan Abbasi, Yassmin Al-Khasawneh, Yuliya Baranovskaya, Marta Besalu, Giulio Brugnaro, Elena Chiridnik, Eva Espuny, Matthias Helmreich, Julian Höll, Shim Karmin, Georgi Kazlachev, Sebastian Kröner, Vangel Kukov, David Leon, Amanda Moore, Paul Poinet, Emily Scoones, Djordje Stanojevic, Andrei Stoiculescu, Kenryo Takahashi, Maria Yablonina and support of Michael Preisack and Michael Tondera. In collaboration with: Institute of Evolution and Ecology, Evolutionary Biology of Invertebrates, University of Tübingen—Prof. Oliver Betz, Department of Geosciences, Palaeontology of Invertebrates, University of Tübingen—Prof. James Nebelsick, Module Bionics of Animal Constructions, University of Tübingen: Gerald Buck, Michael Münster, Valentin Grau, Anne Buhl, Markus Maisch, Matthias Loose, Irene Viola Baumann, Carina Meiser, ANKA/Institute for Photon Science and Synchrotron Radiation, Karlsruhe Institute of Technology (KIT)—Dr. Thomas van de Kamp, Tomy dos Santos Rolo, Prof. Dr. Tilo Baumbach, Institute for Machine Tools, Universität Stuttgart—Dr.-Ing. Thomas Stehle, Rolf Bauer, Michael Reichersdörfer, Institute of Textile Technology and Process Engineering ITV Denkendorf—Dr. Markus Milwich. Funding: Competence Network Biomimetics, KUKA Roboter GmbH, SGL Group, Sika, AFBW—Allianz Faserbasierte Werkstoffe Baden-Württemberg.

ICD/ITKE Research Pavilion 2015-16: ICD Institute for Computational Design—Prof. Achim Menges, ITKE Institute of Building Structures and Structural Design—Prof. Jan Knippers; Scientific Development: Simon Bechert, Oliver David Krieg, Tobias Schwinn, Daniel Sonntag; Concept Development, System Development, Fabrication & Construction: Martin Alvarez, Jan Brütting, Sean Campbell, Mariia Chumak, Hojoong Chung, Joshua Few, Eliane Herter, Rebecca Jaroszewski, Ting-Chun Kao, Dongil Kim, Kuan-Ting Lai, Seojoo Lee, Riccardo Manitta, Erik Martinez, Artyom Maxim, Masih Imani Nia, Andres Obregon, Luigi Olivieri, Thu Nguyen Phuoc, Giuseppe Pultrone, Jasmin Sadegh, Jenny Shen, Michael Sveiven, Julian Wengzinek, and Alexander Wolkow; With the support of Long Nguyen, Michael Preisack, and Lauren Vasey; In collaboration with: Department of Evolutionary Biology of Invertebrates—Prof. Oliver Betz, Departement of Palaeontology of Invertebrates—Prof. James Nebelsick, University of Tuebingen; Supported by: Deutsche Forschungsgemeinschaft (DFG), GETTYLAB and BW-Bank Edelrid, Frank Brunnet GmbH, Forst BW, Groz-Beckert KG, Guetermann GmbH, Hess & Co. KUKA Roboter GmbH, Mehler Texnologies GmbH.

References

1. Andreani S (2014) Over the material, past the digital: back to cities. In: Tedeschi A (ed) AAD–Algorithms Aided Design. Le Penseur, Brienza, pp 343–351
2. Apolinarska A.A, Bärtschi R, Furrer R, Gramazio F, Kohler M (2016) Mastering the "Sequential Roof"-Computational Methods for Integrating Design, Structural Analysis, and Robotic Fabrication. In: Adriaenssens S, Gramazio F, Kohler M, Menges A, Pauly M (eds) Advances in Architectural Geometry 2016, vdf Hochschulverlag AG an der ETH Zürich, pp 240–258
3. Bechthold M (2009) New stone shells: design and robotic fabrication. In: Proceedings of the international association for shell and spatial structures (IASS) symposium 2009, Valencia, 28 September–2 October 2009, Universidad Politecnica de Valencia, Spain, pp 1780–1789
4. Bechthold M (2010) The return of the future a second go at robotic construction. AD Archit Des 80:116–121
5. Bechthold M (2012) Design Robotics. In: Peters B, Peters T (eds) Inside Smartgeometry: expanding the architectural possibilities of computational design, AD Smart. Wiley, Chichester, pp 254–264
6. Bechthold M, King N (2013) Design Robotics. In: Brell-Çokcan S, Braumann J (eds) Robotic fabrication in architecture, art, and design. Springer, Wien, New York, pp 118–129
7. Bier HH, Mostafavi S (2016) Robotic building as physically built robotic environments and robotically supported building processes. In: Schnädelbach H, Wiberg M, Varoudis T (eds) Dalton S. Architecture and interaction, Springer, pp 253–271
8. Bock T, Langenberg S (2014) Changing building sites. AD Archit Des 229:88–98
9. Bonwetsch T (2013) Designing Robotic Assemblies. In: Peters B, Peters T (eds) Inside Smartgeometry: expanding the architectural possibilities of computational design, AD Smart. Wiley, Chichester, pp 218–230

10. Carpo M (2014) Mario Carpo in conversation with Matthias Kohler. In: Gramazio F, Kohler M, Langenberg S (eds) Fabricate: negotiating design & making. Gta-Verl, Zurich, pp 12–21
11. Colletti M (2016) Trans – Disciplinarity. AD Architectural Design, Vol. 243, pp. 77-83
12. De Fusco R (2002) Storia del design. Laterza, Bari
13. Deutsch R (2015) Data-driven design and construction: 25 strategies for capturing, analyzing and applying building data. Wiley, Hoboken
14. Frazer J (2015) Parametric computation. AD Archit Des 240:18–24
15. Garcia MJ, Retsin J (2015) Design methods for large scale printing. In: Real-time, proceedings of eCAADe 33rd annual conference 16th-18th September, Österreichischer Kunst-und Kulturverl, Vienna, pp 331–339
16. Gramazio F, Kohler M (2008) Digital materiality in architecture. Lars Muller Publisher, Zurich
17. Gramazio F, Kohler M, Oesterle S (2010) Encoding materials. AD Archit Des 80:108–115
18. Gramazio F, Kohler M, Willmann J (2014) The robotic touch: how Robots change architecture. Park Books, London
19. Gramazio F, Kohler M (2014) Made by robots. AD architectural design. Wiley, London
20. Grobman Y, Neuman E (2012) Performalism: form and performance in digital architecture. Routledge, London and New York
21. Hansmeyer M, Dillenburger B (2013) Mesh grammars-procedural articulation of form. In: Stouffs R, Janssen P, Roudavski S, Tunçer B (eds) Open systems: proceedings of the 18th international conference on computer-aided architectural design research in Asia (CAADRIA 2013), The association for computer-aided architectural design research in Asia (CAADRIA), Hong Kong, and center for advanced studies in architecture (CASA), Department of Architecture-NUS, Singapore, pp 821–829
22. Heijne R, Van Zwol J, Leupen B (2005) Time-based architecture. 010 Publisher, Rotterdam
23. Hensel MU, Menges A (2006) Morpho-ecologies. Architectural Association, London
24. Hensel MU (2013) Performance-oriented architecture: rethinking architectural design and the built environment. Wiley, Chichester
25. Knippers J, Gabler M, La Magna R, Waimer F, Menges A, Reichert S, Schwinn T (2012) From nature to fabrication: biomimetic design. principles for the production of complex spatial structures. In: Hesselgren L, Sharma S, Wallner J, Baldassini N, Bompas P, Raynaud J (eds) Advances in Architectural geometry 2014. Springer, Wien, NewYork, pp 91–106
26. Kolarevic B (2003) Architecture in the digital age. Design and manufacturing. Spon Press, New York
27. Kolarevic B, Malkawi A (2005) Performative architecture: beyond instrumentality. Spon Press, New York
28. Kolarevic B (2015) From mass customisation to design 'Democratisation. AD Archit Des 238:48–54
29. Krieg OD, Dierichs K, Reichert S, Menges A (2012) Performative architectural morphology finger-joint plate structures integrating robotic manufacturing, biological principles and location-specific requirements. In: Gengnagel C, Kilian A, Palz N, Scheurer F (eds) Computational design modeling, proceedings of the design modeling symposium Berlin 2011, pp 259–266
30. Leatherbarrow D (2009) Architecture Oriented Otherwise. Princeton Architectural Press
31. Lynn G (1998) Folds, bodies & blobs: collected essays. Books-by-architects, La Lettre volée, Bruxelles
32. Menges A (2012) Material Resourcefulness: activating material information in architectural design. AD Archit Des 216:34–44
33. Menges A (2012) Morphospaces of robotic fabrication. In: Brell-Çokcan S, Braumann J (eds) Robotic fabrication in architecture, art, and design. Springer, Wien, New York, pp 28–47
34. Menges A, DeLanda M (2015) The New Materiality. Archit Des 85(5):16–21
35. Menges A, Knippers J (2015) Fibrous Tectonics. AD Archit Des 237:40–47
36. Menges A (2015) The new cyber-physical making in architecture. AD Archit Des 237:40–47
37. Menges A (2015) Material synthesis. AD architectural design. Wiley, London

38. Menges A, Schwinn T, Krieg OD (2016) Advancing wood architecture: a computational approach. Routledge, New York
39. Mogas Soldevila L, Duro-Royo J, Oxman N (2014) Water-based robotic fabrication. J 3D Print Addit Manuf 1:141–151
40. Mogas Soldevila L, Duro-Royo J, Oxman N (2015) Form follows flow: a material-driven computational workflow for digital fabrication of large-scale hierarchically structured objects. In: Combs L, Perry C (eds) Proceedings of the ACADIA 2015 conference–computational ecologies: design in the anthropocene, Cincinnati, Ohio
41. Mogas Soldevila L, Oxman N (2015) Water-based engineering & fabrication: large-scale additive manufacturing of biomaterials. In: Adaptive architecture and programmable matter symposium: next generation building skins and systems from nano to macro, MRS proceedings, vol 1800. Cambridge University Press
42. Mollica Z, Self M (2016) Tree Fork Truss-geometric strategies for exploiting inherent material form. In: Adriaenssens S, Gramazio F, Kohler M, Menges A, Pauly M (eds) Advances in architectural geometry 2016, vdf Hochschulverlag AG an der ETH Zürich, pp 138–153
43. Oxman R (2009) Performative design: a performance-based model of digital architectural design. Environ Plan 36:1026–1037
44. Oxman N, Keating S, Tsai E (2011) Functionally graded rapid prototyping. Taylor & Francis
45. Oxman N (2012) Programming matter. AD Archit Des 216:88–95
46. Oxman N, Kayser M, Laucks J, Firstenberg M (2013) Robotically controlled fiber-based manufacturing as case study for biomimetic fabrication green design, materials and manufacturing processes, Taylor & Francis
47. Oxman N, Dikovski D, Belocon B, Carter WC (2014) Gemini: engaging experiential and feature scales through multi-material digital design fabrication. J 3D Print Addit Manuf 1:108–114
48. Oxman N (2015) Templating and design biology for biology design. AD Archit Des 237:101–107
49. Picon A (2014) Robots and architecture: experiments, fiction, epistemology. AD Archit Des 229:54–60
50. Retsin J (2016) Discrete assembly and digital materials in architecture. In: Complexity & simplicity, proceedings of eCAADe 34rd annual conference 22th-26th August, pp 143–151
51. Schumacher P (2015) Parametricism 2.0. AD Archit Des 240:8–13
52. Scheurer F, Schindler C, Braach M (2005) From design to production: three complex structures materialised in wood. In: 6th international conference generative art 2005, Celestino Soddu, Milan
53. Sheil B (2012) Manufacturing the Bespoke. AD Reader. Wiley, United Kingdom
54. Wonderlab (2016) Vision. http://www.w-o-n-d-e-r-l-a-b.com/vision/. Accessed 24 June 2016
55. Willmann J, Gramazio F, Kohler M, Langenberg S (2012) Digital by Material. In: Brell-Çokcan S, Braumann J (eds) Robotic fabrication in architecture, art, and design. Springer, Wien, New York, pp 12–27
56. Yuan P (2015) Parametric regionalism. AD Archit Des 240:92–100

Chapter 2
Informed Architecture and Clay Materials. Overview of the Main European Research Paths

Abstract The chapter proposes a reflection on the role of technological modernization in the reintroduction of ancient construction techniques and traditional materials, commonly used in the Mediterranean. The continuous flow of traditions and innovations has affected architecture through the introduction of materials and construction technologies for the development of dwellings and later of cities and infrastructures. Stone, clay, as well as wood, have for a long time represented the constituent elements of an architecture shaped by places and cultures. The advent of modernity has radically transformed this relationship by bringing architecture back to a partial homogenization driven by mass production of building components. In the Anthropocene era, the diffusion of information technology processes and modern manufacturing technologies is a stimulus for the reintroduction of traditional materials and techniques within a new tectonic vision to recreate the correlation between project, material, and process production which was a characterizing element of Mediterranean architecture. The analysis focuses on the relationship between computational design, digital fabrication, and clay in the creation of informed technological systems in order to demonstrate how it is possible to reinterpret ancient knowledge through innovative technologies.

Keywords Clay design · Parametric design · Fabrication strategies · Informed technological systems · Complex clay structures · Optimization of clay technological devices.

2.1 Introduction: Architectures in Raw Earth in the Mediterranean

Although we can trace the traits of environmental and climatic similarities, the "great sea", as the Mediterranean Sea is called, has always been the cradle of different cultures and traditions, which come together in this boundless territory [1]. The cultural interference between people is testified through the artistic and architectural heritage that has survived to the present day, which enables humanity to "meet the Roman world in Lebanon, prehistory in Sardinia, the Greek cities in Sicily, the Arab

A. Figliola and A. Battisti, *Post-industrial Robotics*,
https://doi.org/10.1007/978-981-15-5278-6_2

presence in Spain, the ‘Turkish Islam in Yugoslavia” [4], as a tangible and concrete manifestation of these dynamic flows. Among the materials and construction technologies that have become more widespread, clay stands out. Starting from the first Neolithic civilizations and the birth of Jericho in 8000 BC, the first stationary city that sanctioned the end of the nomadic status of populations, this element was used in fabrication procedures and in time established itself in all areas of economic and cultural development [7]. Raw earth architecture has long represented a sustainable, economic, and technological solution characterized by relative simplicity in terms of manufacturing and installation of components, as well as applicability in different climatic areas. The diffusion of this kind of material and of the related production technologies to process it, among which adobe construction, pies, and forming technique [1], are testified by the discoveries made and the fact that this technology is still used for the building of artifacts in various regions of the Mediterranean [9] and Central America [5] (Fig. 2.1). In addition to the above-listed technical characteristics, the production process pertaining raw earth architecture constructions defines the figure of man as an active subject able to play a fundamental role in place shaping dynamics [11], and creating a mutual bond between performance, material, climate, and cultural traditions. With the advent of modernity and industrial culture, the symbiotic relationship between form, function, and environment has disappeared giving way to the homologation of the three elements mentioned above. Form follows function, a slogan of modernity presented by Luis Sullivan,

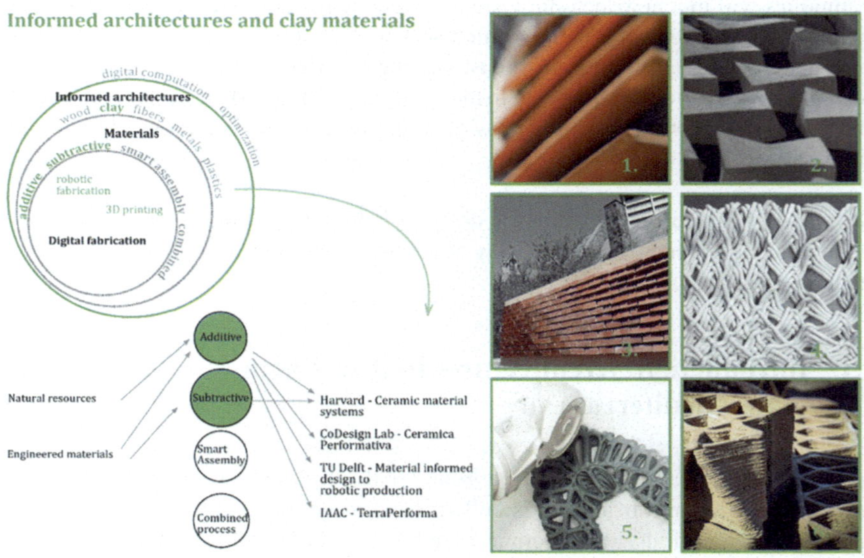

Fig. 2.1 Informed architectures and clay materials. Overview of the main research topics. © Angelo Figliola

represented the underlying design paradigm that allowed to "adapt an idea of homogeneous and isotropic space to an inhomogeneous and anisotropic energy field" [3]. Based on this new example, the morphological considerations underlying vernacular Mediterranean architecture—capable of adapting to different climatic conditions through the synergy between material distribution, spatiality, and microclimate [8]—are replaced by mechanical principles that tend to standardize the behavior of buildings adapting to changing climates. In the historical recourses of architecture, the response and adaptation to non-homogeneous energy fields occurred through passive devices whose morphology was generated based on the energy streams following the precept of "Form follows flow", or exploiting the performance characteristics of materials. Examples of this design approach are the traditional clay structures in Mediterranean countries (Fig. 2.2), where the relationship between interior space and external atmosphere has stimulated the creative process and fostered the invention of passive technological systems as massive devices for the building envelope, wind towers, sunscreens, as well as the study of spatial configuration aimed at increasing indoor comfort conditions [12]. The advent of mechanical devices has accelerated what the industrial revolution had initiated, shifting the design focus from form and materials to mechanical devices, while reducing the complexity of the spatial organization. In the era of Anthropocene [6], the growing need to cope with environmental

Fig. 2.2 Ceramic shading systems, 1:1 scale prototype. @ Martin Bechthold

problems requires the development of alternative design solutions capable of reintroducing traditional materials, technologies, and design processes using the mechanics that characterize the industry 4.0 [2].

2.2 Performance-Based Architectures and Adaptive Manufacturing

The integration between computational processes and innovative fabrication methods allows the reintroduction of natural and traditional materials, such as wood, clay, and stone, in the design and construction procedure [10]. This approach represents the basis for exploring the correlation between shape generation and performance evaluation according to design criteria related to the energetic or structural behaviors of the morphologies. Among the components quoted above, another difference should be made to better identify the nature of the source. In the case of clays, it is possible to utilize two different kinds of materials: natural clay and ceramic derived from industrial processes. Their production and post-production phase distinguishes the two materials: the fabrication of ceramic components needs a drying and cooking procedure employing high-temperature ovens, while natural clay can be applied directly after a drying process without using expensive ovens and complex procedures. Clay presents a series of advantages in architecture and engineering characterized by its excellent mechanical and structural properties, as well as by being a biodegradable, low cost, and zero-kilometer material, as it can be sourced directly on site. As a construction component, clay allows the reduction of energy loads related to the building's heating and cooling demands thanks to their thermal inertia. Furthermore, clay technological administration act as adaptive devices, adjusting comfort conditions through absorption and evaporation processes, controlling the percentage of humidity present in the environment [23]. Apart from the features of the element described above, another important stimulus for research on the implementation of technological clays systems in architecture was represented by innovations in the field of digital fabrication and construction. The simultaneous affirmation of 3D printing and robotic manufacturing has favored the birth of a hybrid research path, combining the potentialities of both technologies. This highly performative approach can be defined as Large-Scale Robotic 3D printing, or Robotic for Additive Manufacturing (RAM). More specifically, combining the two manufacturing technologies mentioned above, it is possible to explore the pros deriving from the genericity of the robot as an industrial machine applied to additive processes. The generic nature of the robots allows establishing a customized building procedure, experimenting with different morphological configurations and various material systems.

The customization of the production phase is realizable as a result of the following technical characteristics of the machine:

- 6° of freedom instead of the 3° of a common 3D printing machine;

- Possibility to design and prototype different end effectors according to a specific fabrication process;
- A large working area if compared with other digital manufacturing tools available on the market.

The potential to move with such degree of freedom and to connect the fabrication phase with the digital design through constant models represents a big opportunity to overcome the size constraints of traditional 3D printing technology, optimize the material distribution, and reduce waste production through the limitation of the support and infill secondary structures. Programing robots with parametric software allows linking shape generation with the manufacturing process through a design-to-production path. The implementation of computational logic into the design procedure represents the main conductor to inform the models with performative criteria related to energy and structural behavior. The concept of informed morphologies connected to a large-scale robotic 3D printing technique opens an entirely new field of investigation on innovative clay material systems and fulfilling technological devices. In regard to production technologies, the experiments mainly involve two processes: additive manufacturing, the smart assembly of discretized components, and subtractive processes such as NC wire cutting through which the diminishing of material defines the shape of the component. Both procedures embed the modernization from traditional production processes: reductive technologies represent the reinterpretation of clay molding procedures at the base of the technique known as adobe, carried out by creating wooden casts; 3D printing avoids the creation of support and beating structures, typical of the pisè technique, simply by controlling the extrusion pressure during material deposition.

2.3 Research Lines in the Field of Performative Clay

To explore this field of the investigation a series of applied research activities are being conducted among the advanced laboratories in the European universities and research centers. The project *Pylos*[1] developed by the Institute for Advanced Architecture of Catalunya (IAAC) is the starting point of the research line on clay in architecture. The project is based on the use of biodegradable and reusable clay through an innovative manufacturing process. The application of this kind of material possessing which has excellent mechanical and thermal qualities in the design process allows building performative and large structures with a maximum height of 2.2 m without the time-consuming proceeding of drying and baking. The research aimed at building printed clay columns to fully explore the potential of shape customization according to performative criteria, and the technical advantages of the robotic production procedure. Implementing manufacturing processes utilizing

[1]IAAC (2017), *Large Scale 3D printing, Pylos*, available at: https://iaac.net/research-projects/largescale-3D-printing/pylos/ (accessed 26 April 2017).

robots fosters the evaluation on modern production methods and traditional materials ensuring precision and surface quality, which are not possible with traditional techniques. Based on the results of the *Pylos* research project, IAAC continues to examine the correlation between performance-based morphologies and robotic additive manufacturing through the Open Thesis Fabrication (OTF) research program. During the OTF 2016/2017 an experimental 1:1 scale prototype of building envelope made of natural and unbaked clay was assembled. The research concerned the study of optimized geometrical patterns according to energy and structural presentation and the analysis of the basic technological module to ensure an easy and fast assembly process directly on site. As a result of the computational procedure and the shape optimization a visually static informed morphology was defined. Furthermore, the complex pattern of the external surface was maximized according to values of direct solar radiation and natural ventilation: the design methodology adopted permitted the reduction of the surface temperature of the component through self-shading, consequently decreasing the energy load needed for the cooling procedure of a building. Another research line on the topic of clay and informed morphologies is offered by the Harvard GSD Graduate School of Design, and the Robotic Design Group (RDG) directed by Professor Martin Bechtold [13]. The examination took into account the relationship between academic research and industrial applications. The RDG conducted studies on the object of discussion with a pragmatic approach, analyzing traditional manufacturing processes and consequently identifying which phase of the production can be implemented using innovative technologies. The focus of the research was on the following:

- Informed Technological Systems Concerning Energy and Structural Performative Criteria
- Implementation of robots into the standard manufacturing process [15].

A process of innovation is needed to explore new informed formal codes and to expand the range of design possibilities. The evaluation regards the software and hardware components. For the software component, the studies focus on the design-to-production process and aim at making the tools available and the computational workflow powerful and easy to use. The customization of the production procedure has also boosted the studies on the hardware part, introducing the design and prototyping of end effectors according to different manufacturing processes and material systems. The automation of procedures and the increase in fabrication efficiency is sided by an exploration of the potential offered by introducing robotic construction together with a generative computational process, whereby the performances become design inputs within a logic of iteration and data-driven procedure. The research strategy and the relationship with the industrial sector are based on a real protocol that provides the following steps:

- Analysis of the type of architecture and the technological and material system;
- Examination of the technical aspects of the fabrication procedure;
- Evaluation of the production, distribution, and economic process;
- Summary of exploration and formulation of the potential/critical framework;

- Definition of possible implementations of the design and manufacturing processes;
- Design of a material system founded on the customization of a basic module and the aggregation of the latter.

Based on the study protocol described above, the RDG research group explored the association with the world of industrial production through collaboration with the Spanish industry "ASCER Tile of Spain" on the customization and manufacturing of ceramic systems. The research collaboration concerns the need to introduce degrees of freedom in the construction process to ensure greater flexibility. Being able to set up flexible fabrication procedures ensures the possibility to customize the morphologies according to the performance parameters used as a guide of the design process and, at the same time, guarantees economy and sustainability of production procedures. The studies were conducted with a holistic, multidisciplinary approach, tackling the problem from all points of view: from the ceramic manufacturing process up to the installation and possible dismantling, considering the optimization of material resources and the reduction of waste. Some research projects were a result of the cooperation between academia and industry, and they led to moderation of the fabrication process due to the introduction of robotic manufacture. The latter was aimed at product customization and material resource maximization without upsetting the entire production process but working instead on the innovation of specific procedures within the building chain. Robotic manufacturing applied to architecture represents another research path pursued by the TU Delft Hyperbody Research Group [21]. The focus of the experiments, defined as robotic building,[2] is the integration of robotic production into the entire design and construction process—from modeling to simulation, up to manufacturing—aimed at creating visually static informed architectures. In this type of process, performance becomes the guiding parameter of the formal generation procedure, and the resulting tectonic complexity can be materialized through new digital fabrication technologies which guarantee greater flexibility in production processes. Taking advantage of the technical characteristics of the tools employed, various fabrication methods were investigated, including the 3D printing additive procedure utilizing plastic polymers or clay. Through an integrated process, design-to-production, it is possible to realize structures with the following traits:

- Customized and diversified in terms of material and production procedure;
- Energetically efficient;
- Structurally performative;
- Morphologically responsive.

[2]Robotic Building is a three year research program (2014–2018) of the Hyperbody Research Group at the Technical University of Delft. The investigations are conducted by Henriette Bier as project leader, Sina Mostavi, Ana Anton and Serban Bodea as researchers. Collaborate in research 3TU Bouw, ABB Robotic, KUKA, 100% TUD, Delft Robotics Institute.

2.4 Performative Clay: Case Studies

To review the different techniques, the book proposes a series of case studies and the results of applied research (Fig. 2.1). The case studies are selected based on their correspondence to the following parameters:

- Presence of a performance-based process to explore informed architectures;
- Use of ceramics or natural clay;
- Materialization of the digital model through innovative manufacturing procedures, specifically robotic fabrication, and 3D printing techniques.

All the projects result from a performance-based process, while performative parameters (e.g., structural, environmental) inform the generative procedure.

2.4.1 Harvard GSD: Integrated Environmental Design and Robotic Fabrication Workflow for Ceramic Shading Systems

The proposed case study is part of a series of experiments carried by the Robotic Design research group[3] at Harvard GSD aimed at examining the new design potentialities that digital computation and modern fabrication methods open in terms of employment of traditional materials, such as clay, in the design of responsive solar shading devices (Fig. 2.2) [13, 15].

The representation of the environment consists of the Rhinoceros Software and the generative modeling plug-in Grasshopper. The latter allows the integrated use of environmental and thermal analysis software such as Diva and Energy Plus (Fig. 2.3). The examination started from an existing architectural volume regarded as a testing room with a large curved glass surface oriented toward the south: in the setting up of thermal exploration, the volume is considered as a single thermal zone where the opaque walls are considered adiabatic, while for the transparent walls the thermal properties are specified. Through the management of geometric parameters and structured data, the model generates different solutions of shading systems that affect the thermal analysis and behavior of space [14]. The evaluation is aimed at finding the optimized geometrical solution to match the objectives: that is, to maximize thermal gain during two inspection periods on an annual basis. The need to achieve multi-objective maximization led to the generation of shielding technological components based on a conic surface. This guarantees the achievement of the previously established performances.

The optimization process is founded on a meta-project tool which can be used by the designer to define the geometrical constraints and parameters (genotype and

[3]Design Robotic Group (2012) Ceramic System. Available at: http://research.gsd.harvard.edu/drg/category/ceramic-systems/custom-systems/ (accessed 29 November 2016).

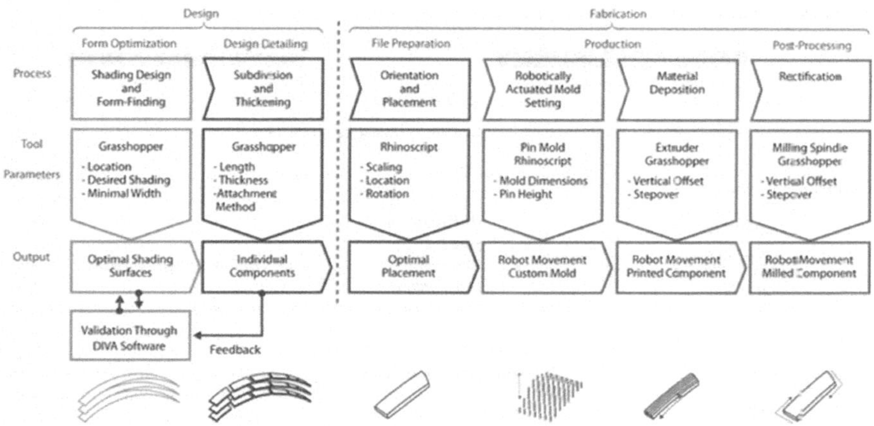

Fig. 2.3 Ceramic shading systems, computational workflow. @ Martin Bechthold

phenotype) according to the value of the thermal evaluation. Subsequently to the analysis, the innovative methodology introduced by the case study explored modern design solutions, in terms of morphology and materiality. The variable model allowed a direct comparison of the results obtained from the examination. The values of energy consumption on an annual basis expressed in kWh/m2 demonstrated the validity of the process and the complexity of the morphology derived from the optimization of the device. As a result of the application of robotic fabrication technology, it was possible to realize a non-standard shape shielding system, a result of the algorithmic optimization, through the standardization of the digital and production procedure rather than of the final result. Following the research protocol of the research group, the use of a robot for clay extrusion and the employment of a dynamic mold modernizes the fabrication process. This procedure is dynamic thanks to digitally controlled metal pins that enable the mold to become suitable for extrusion of free-form components. An environmentally friendly material such as ceramic gives the whole experiment a greater value in terms of environmental sustainability parameters, life cycle, and grey energy. The proposed methodology (Fig. 2.4) can be involved in the realization of shielding systems in new-build multi-layer enclosures or for retrofitting interventions, as well as for technological networks applied to opaque vertical partitions.

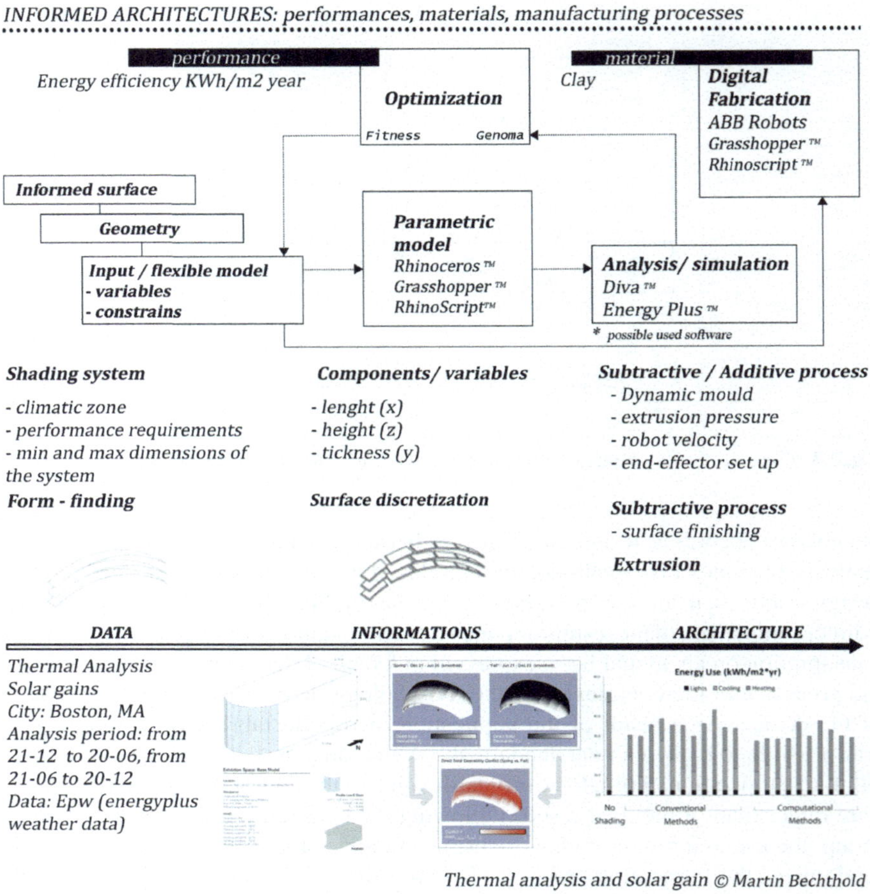

Fig. 2.4 Ceramic Shading System: performances, materials, manufacturing processes. A design methodology. © Angelo Figliola

2.4.2 *Harvard GSD. Flowing Matter*[4]*: Robotic Fabrication of Complex Ceramic Systems*

The case study *Flowing Matter*[5] is one of the experiments operated by the University of Harvard GSD on innovative ceramic complexes, as part of the collaboration between the research center and the Spanish company Ascer Tile of Spain

[4]Design Robotic Group (2013) Ceramic System. Available at: http://research.gsd.harvard.edu/drg/ceramic-systems/flowing-matter/ (accessed 28 June 2016).

[5]Flowing Matter (2012) Generation of complex tectonic ceramic assemblies composed of robotic wire-cut ruled-shaped elements. You Tube Video, 20 February. Available at: www.youtube.com/watch?v=vFOdNKe3kIY (accessed 29 June 2016).

Fig. 2.5 Complex ceramic systems, 1:1 scale prototype of customized bricks. @ Stefano Andreani and Martin Bechthold, Harvard University Graduate School of Design

[15]. The investigation aims to explore new design possibilities offered by digital computing and robotic fabrication when utilizing ceramic materials in the design of performative technological systems and for the realization of non-standard building envelopes (Fig. 2.5) [16]. The research project development originates from the study of consolidated and experimental structural solutions, from the ceramic edifices of Eladio Dieste up to the contemporary Catalan Vault of Philippe Block. The examination, which involves the construction of ceramic components together with the analysis of their industrialized production, has the objective of identifying the new lines of development and modernization within the field. Through the definition of a complex parametric model, it was possible to explore different spatial configurations and to evaluate innovative material and tectonic systems. Likewise, it was possible to increase the system's structural performance, thanks to the expansion of the contact area between the shear-resistant components and ensure easy on-site assembly.

The discretization of double-curved surfaces has led to the definition of a joint network where each element is different from the other. The traditional industrial fabrication chain, where the ceramic material is constructed by extrusion and processing on a linear chain, was implemented to produce the customized components. The introduction of two robotic arms—equipped with a hot metal wire and coupled with the digital procedure of design-to-production—within the traditional supply chain allowed to control the complexity of the shapes by establishing a cutting path through geometric elements such as points and lines. The prototype demonstrated the inexpensive and easy management of the process; a procedure that explores new spatial and structural solutions starting from the study of traditional processes. More specifically, the parametric model permitted the examination of a series of compound geometries able to optimize the structural performance, increasing the shear-resistant contact surface and ensuring an easy and fast assembly procedure, due to perfectly interconnected components derived from the discretization of the surface. Furthermore, through the digital model, it was possible to realize and analyze two different prototypes: a double-curvature surface and a surface with custom-made openings adapted to external inputs. The customized components were attained by executing the ceramic extrusion process of industrial derivation through two robotic arms equipped with a hot wire. The latter can create spatial cuts following coordinates provided by the digital model. The proposed methodology (Fig. 2.6) appears to be easy to manage, low cost, and applicable for the achievement of responsive opaque vertical partitions. Management of the digital model and production procedure are the critical aspects of the proposed process; in fact, it presents numerous variables that should be controlled.

2.4.3 Harvard GSD: [R] Evolving Bricks

The case study *[R] evolving brick*[6] (Fig. 2.7) [17] proposes a technological system applicable to opaque external partitions to modulate the activation of thermal mass through different design strategies and geometric variation processes. The standard geometry of clay bricks (rectangular block) is parametrically modified through the employment of two different design strategies.

- Single cut managed by a spline curve and a ruled surface;
- Double cut realized by rotated planes lead through the rotation angle.

The project follows a research path initiated by tectonic and spatial experiments with clay components at the ETH Zurich Polytechnic, by the research group led by architects Fabio Gramazio and Mathias Kohler. Unlike formal trials, the case study introduces new design data related to technological and environmental performance. The geometry obtained by a digital computation and fabrication process allows managing solar incidence, ensuring a selective activation of the thermal mass, and reducing the thermal flow between the external and internal environment. By first bringing innovations in the traditional clay fabrication procedure, the research was carried out to obtain an efficient and inexpensive production chain to guarantee high possibilities of topological variation [18]. The adopted manufacturing method involves the introduction of anthropomorphic robots in the traditional production chain without subverting a standard and efficient process. A robotic fabrication protocol manages the subtractive process of shaping the clay blocks with a CNC hot wire cutting procedure, making it possible to manage different cutting

[6]Andreani (2016) Revolving Brick. Available at: http://www.stefanoandreani.com/portfolio/revolving-brick/ (accessed 29 November 2016).

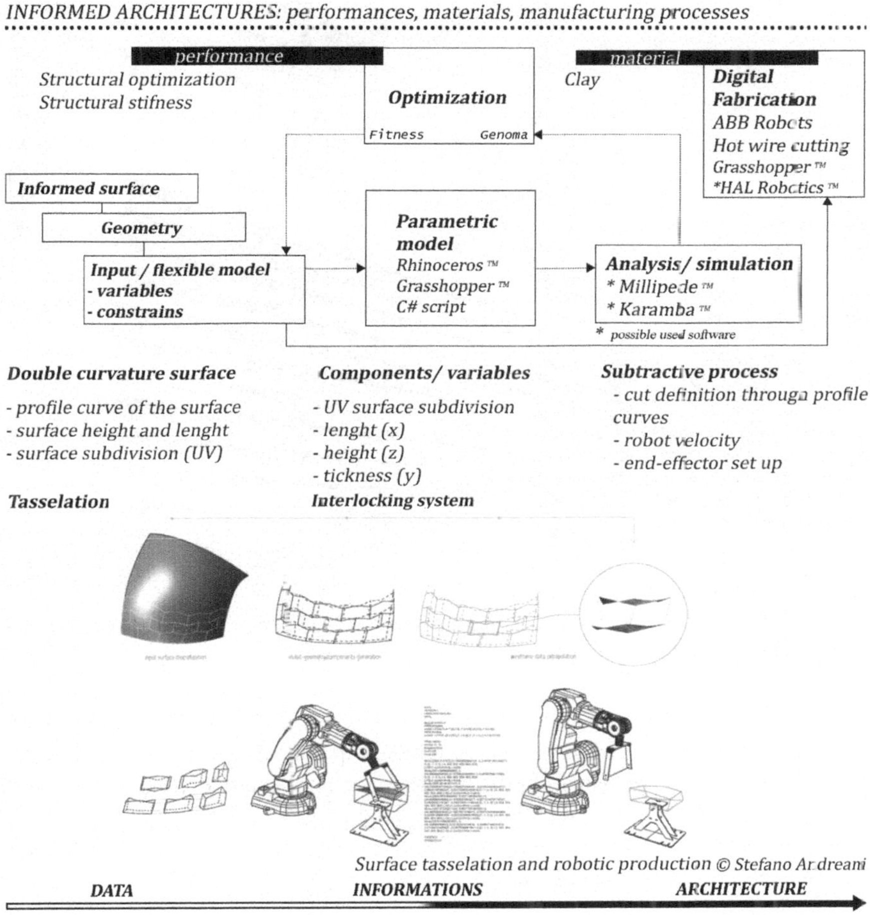

Fig. 2.6 *Flowing Matter*: performances, materials, manufacturing processes. A design methodology. © Angelo Figliola

plans through intense man-machine collaboration. The assembly strategy is also part of the design process as a result of the customization of components provided by the technological system. Furthermore, the parametric procedure is fundamental to elaborate cataloging of elements, which in turn facilitates the operations of on-site construction. The analyzed case study proposes a technological system based on the assembly of customized clay bricks able to selectively modulate the activation of the thermal mass and consequently control the thermal flow between the external and internal environment. The comparison between the proposed prototype and the traditional system made with rectangular blocks showed how the design objective of a significant thermal flow reduction on an hourly basis was achieved. The system's responsiveness, perfectly scalable for applications in opaque vertical partitions, is

Fig. 2.7 *[R] evolving bricks*, 1:1 scale prototype of customized bricks. @ Stefano Andreani and Martin Bechthold, Harvard University Graduate School of Design

linked to the design strategy and the corresponding climatic zone. This offers a wide range of geometric configurations. The production process involves the introduction of small-sized robotic arms within the traditional supply chain to create customized forming operations as opposed to the informed digital model. The complex assembly process is the critical point of the procedure that requires a simplification strategy starting from the 3D model (Fig. 2.8).

2.4.4 Harvard GSD: Woven Clay

The *Woven Clay*[7] research (Fig. 2.9) proposes the realization of a clay technological unit that investigates a traditional formative principle based on the weaving technique using a non-linear and fluid-viscous behavioral material, controlled by an additive manufacturing robotic fabrication process [19]. Through a computational design protocol, the material network is informed by performance data identified as a project target. The control of design variables that identify the customary interweaving system, such as the number of layers and porosity, allows to examine

[7]Flaherty (2015) Harvard Robot Whiz Invents a Way to Weave Facades Out of Clay. Available at: http://www.wired.com/2014/06/harvard-robot-whiz-invents-a-way-to-weave-build-facades-out-ofclay/ (accessed 21 June 2016).

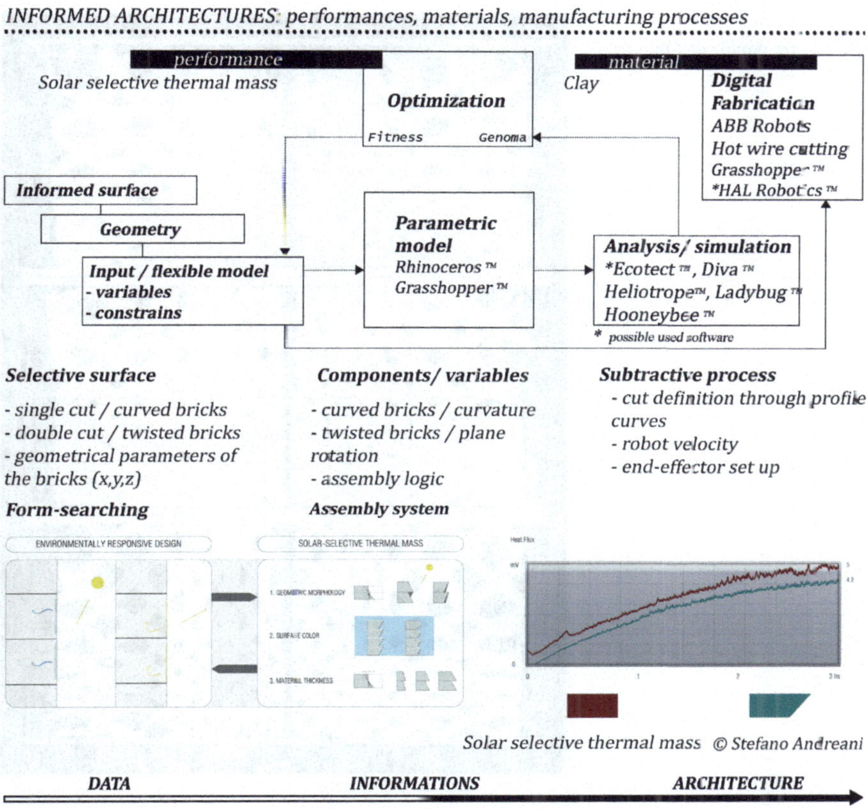

Fig. 2.8 [*R*] *evolving Bricks*: performances, materials, manufacturing processes. A design methodology. © Angelo Figliola

and subsequently verify a series of different morphologies concerning the identified achievements. This specific design approach and the related operative methodology considers the component as a real-design parameter that acts as a design agent. A feedback loop protocol on extrusion variables between analog tests on the material system and the digital one is used to inform the digital process and define the right set up to fabricate the components. Another fluctuating parameter is the definition of the right clay mixture in line with the project objectives: in this case, the evaluation aims at the review of a clay composition whose formal result is plastic and whose variation allows to reach morphological conformations which are difficult to stipulate in the conceptual phase of the project.

The expulsion process was obtained through a robotic arm equipped with an extruder with a nozzle adapted to the thickness of the individual layers. What makes this experiment unique is the use of a 3D mold which performs the extrusion of the

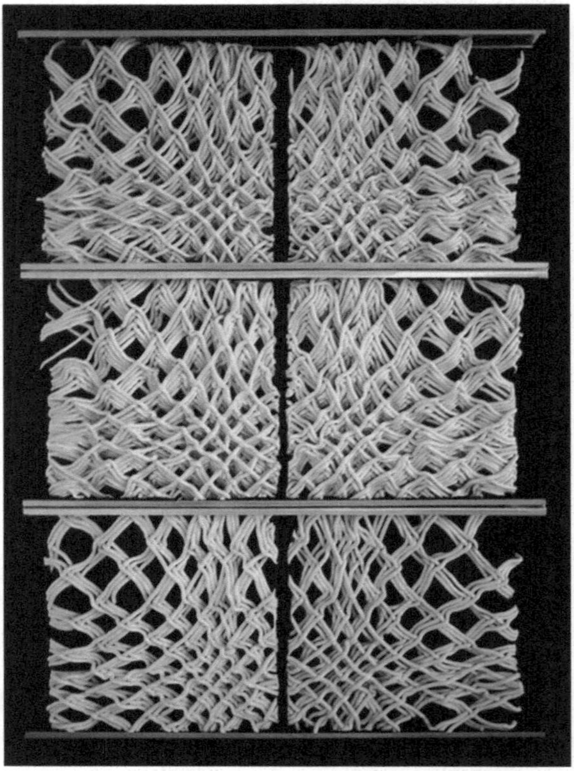

Fig. 2.9 *Woven clay*, 1:1 scale façade panel. @ Olga Mesa

clay, studies its formal result, and identifies any critical issues [20]. Once the technological unit is defined, an aggregation system is studied to verify scalability and applicability as a shielding element for architectural enclosures. The analyzed case study proposes the innovation of a traditional material system (weaving) through a flexible digital model and a robotic manufacturing process, establishing a feedback loop between the virtual and the physical model, considering the material itself as a project variable. The production procedure enables testing with different morphologies according to the desired accomplishments, opening up numerous design possibilities. In order to evaluate its applicability to real-scale architecture and to guarantee the required performances in relation to structural rigidity and resistance to external atmospheric agents, the designed technological system requires further implementations and tests, especially in the definition of the components minimum and maximum dimensions, the printing setting, and the customized technological unit assembly network. A sustainable material together with an efficient fabrication method makes the proposed technological complex suitable for refurbishment operations, as a responsive shielding element in new construction of multi-layer vertical partitions, or in acoustic applications for indoor environments (Fig. 2.10).

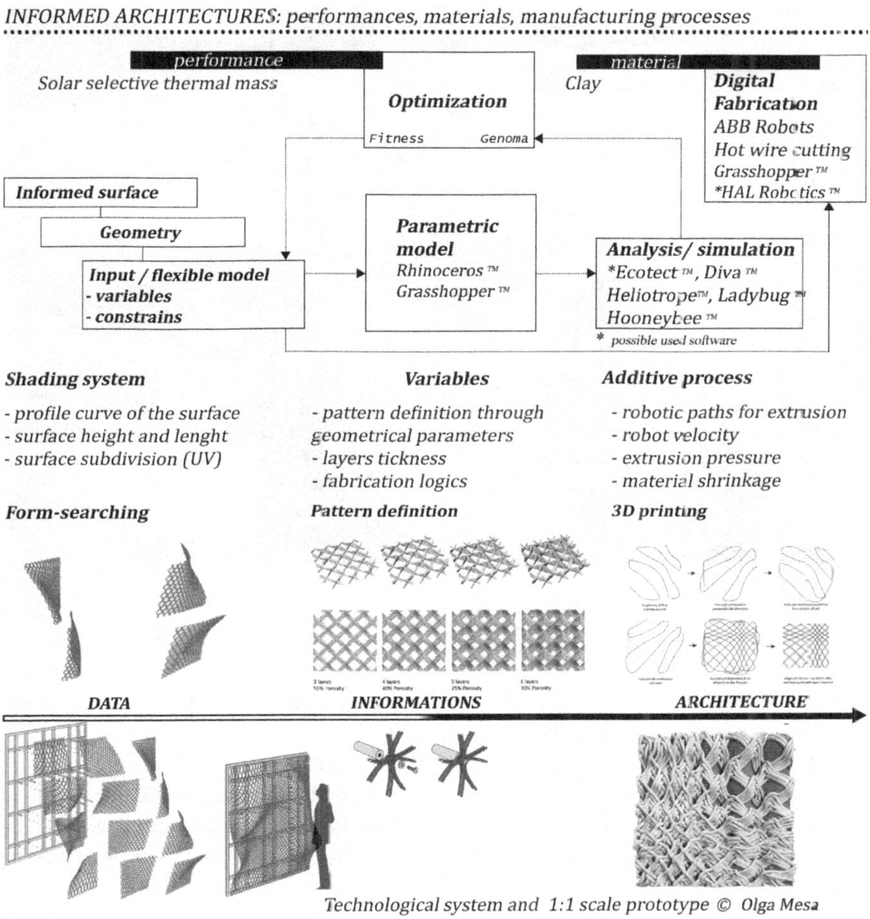

Fig. 2.10 *Woven Clay*: performances, materials, manufacturing processes. A design methodology. © Angelo Figliola

2.4.5 IAAC: Digital Adobe, TerraPerforma

The combination of a natural material such as clay and the additive processes of robotic 3D printing is the basis of some research projects generated at the IAAC aimed at exploring the potential of raw earth for construction of massive technological systems destined for the architectural field. The first experimentation concerning the issue is the *Pylos* project developed at the IAAC and directed at testing the real potentials of robotic and raw earth manufacturing in building, verifying the structural performance qualities of the proposed material system. Based on the knowledge gained with the *Pylos* project, a further step for the research path consists in the

Fig. 2.11 *TerraPerforma*, 1:1 scale of customized bricks. @ IAAC, Institute for Advanced Architecture

Digital Adobe, the construction of the *Terra Performa* prototype (Fig. 2.11), realized within the Open Thesis Fabrication 2016/2017.

The 1:1 scale prototype of a massive soil envelope allowed exploring the possibilities offered by technological innovations to reinterpret technological systems in natural clay common in the Mediterranean and verify their employment in a 1:1 scale through the assembly of technological units that can be easily moved on site. The pioneering research involved in the first phase the definition of the right material mix capable of guaranteeing the optimal structural performance for future engineering applications. After establishing the correct mixture between clay, water, and gelatinous proteins, it was necessary to set up a utensil to extrude the material while checking the variable criteria, such as speed and pressure, responsible for the correct deposition of the component. Parametric tools represented the medium between the formal generation process and the precision of the parameters that inform the production procedure—such as the material deposition path—within a design-to-production logic. Tests have shown that the material possesses excellent mechanical qualities to construct structural systems following a process of customization of the geometries (Fig. 2.12) and fabrication criteria. In the second phase of the research, a series of geometric patterns were produced and analyzed to increase the structural and energy execution of the façade system. The enhancement procedure generated a morphology that guarantees optimal performance due to specific geometric shapes. Likewise, the optimization of the geometric pattern concerning quantitative criteria, such as solar radiation and natural ventilation, significantly reduces the component's

Fig. 2.12 Robotic additive manufacturing of customized bricks. @ IAAC, Institute for Advanced Architecture

surface temperature, through self-shadowing, and thus decreases the energy load required for summer cooling. The energy-environmental performance traditionally assured by the massive wrappers obtained through the adobe or pisè technique is preserved by filling the cavities with solid elements, like sand or soil found on site (Fig. 2.13).

2.4.6 TU Delft, Hyperbody: Materially Informed Design to Robotic Production

The proposed case study *Materially Informed Design to Robotic Production* [21] is part of a research line that examines the phenomenon of porosity for the realization of informed constructions resulting from a process of structural maximization and produced through robotic fabrication. The construction of the 1:1 scale prototype (Fig. 2.14) is the result of a consequential procedure that involved different project levels, from macro- to meso- and microscale, with the application of diverse methods such as light painting, pattern study, and clay printing. At the macroscale, a form-finding process generated the mesh; at the end of this proceeding, the part that should be built as a 1:1 scale prototype was identified. At the microscale level, the focus was on the definition of the material system and the geometric pattern, both defined by following a structural optimization procedure. The intersection of stress lines obtained through structural analysis and horizontal planes created by a series of

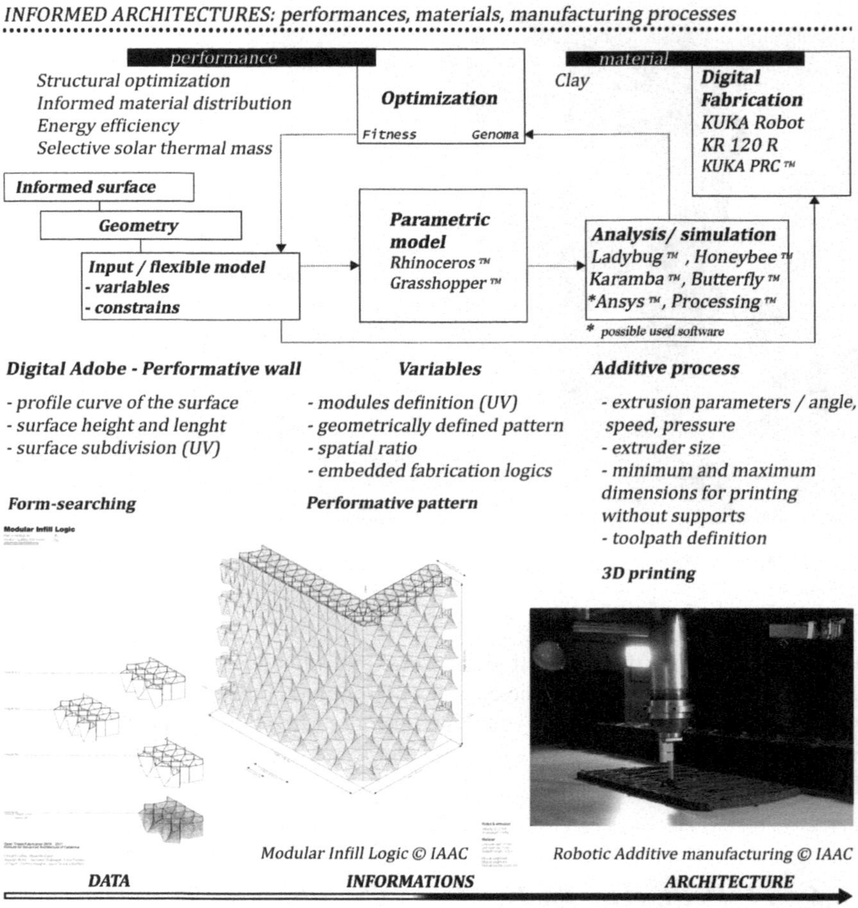

Fig. 2.13 *Terraperforma*: performances, materials, manufacturing processes. A design methodology. © Angelo Figliola

vertical sections of the component creates a series of points that make up the pattern, imitating the Voronoi diagram geometric logic.

The developed pattern distributes the material where it is needed to guarantee the required structural performance, creating a series of variations in terms of density and porosity parameters. Parallel to the evolution of the computational maximization procedure at the microscale, a customized robotic end effector for the extrusion of ceramic fluid-dense material was designed. Tests were carried out on the component system to find the right mixture (e.g., clay + EPS, clay + fibers) to ensure density and plasticity, as well as stability, during the extrusion process [22]. The case study

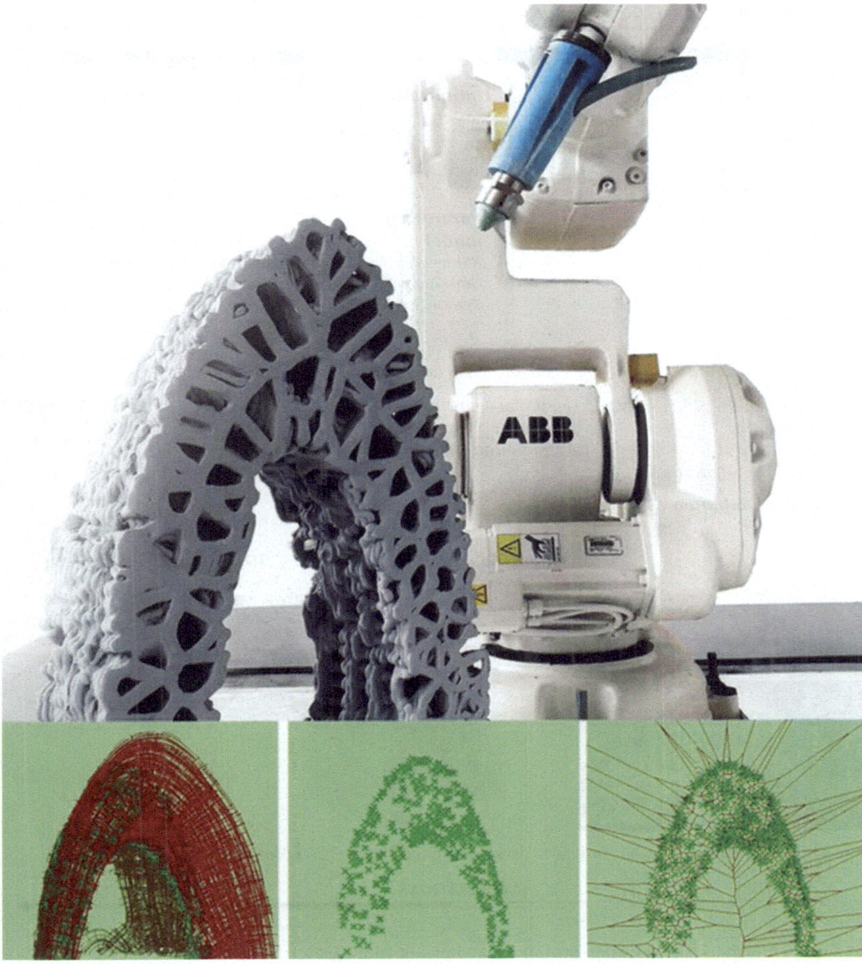

Fig. 2.14 *Materially Informed Design to Robotic Production.* @ Sina Mostafavi

investigates the phenomenon of porosity as a result of the relationship between solid and void on the basis of different scales of studies:

- Macroscale, architectural organism;
- Mesoscale, technological system;
- Microscale, material system.

The 1:1 scale prototype represents the result of the application of the methodology described above: an informed surface able to optimize material and structural executions based on their programatic function. The structural maximization follows all

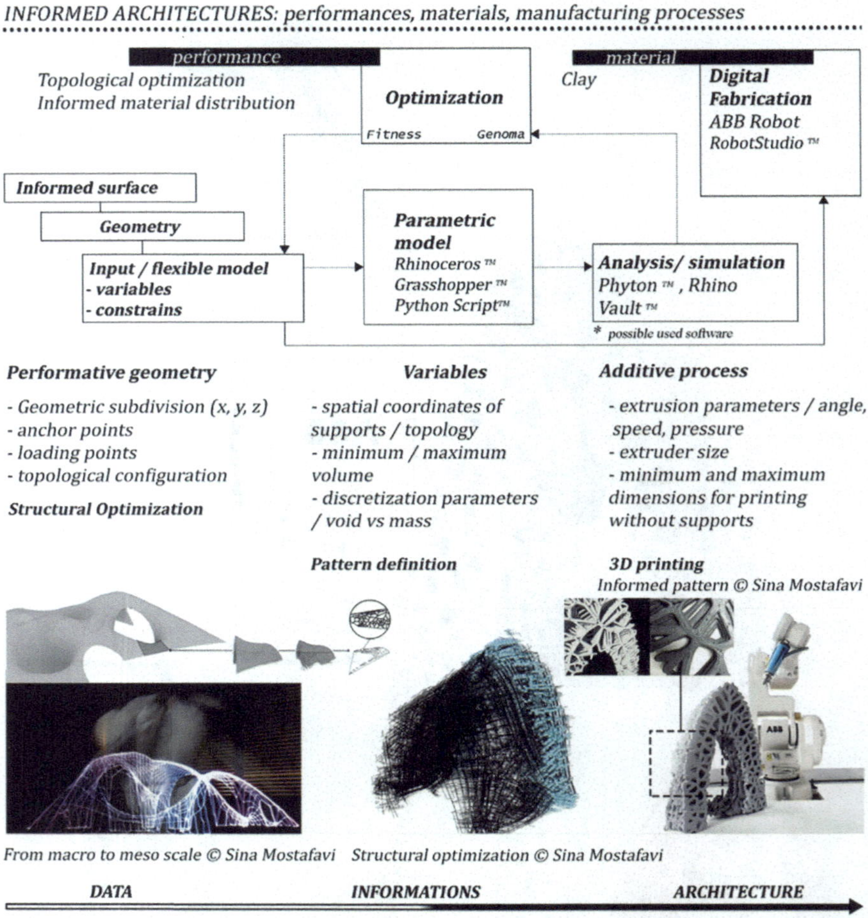

Fig. 2.15 *Informed Porosity*: performances, materials, manufacturing processes. A design methodology. © Angelo Figliola

the phases of the project and contributes to the determination of a performative material system whose customization is guaranteed by the digital fabrication process. The proposed methodology can be applied to the realization of non-standard constructional components and for the fabrication of optimized vertical partitions in which the correlation between solids and voids and the resulting volume can become a planning opportunity. The dimensions of the robotic arm and the geometric coordinates that delineate the working space can be seen as a critical aspect of the proposed methodology (Fig. 2.15) and the material system utilized. A possible implementation of the methodology can derive from the development of a feedback loop network between the digital model and physical prototype, able to adapt manufacturing to the real environmental conditions.

Fig. 2.16 *Ceramica Performativa*, 1:1 scale prototype. @ Paolo Cascone

2.4.7 CODESIGNLAB. Ceramica Performativa

Ceramica Performativa[8] experiments 3D printing of fluid-dense clayey materials for the design of a responsive technological system for newly built architectural façades or retrofitting operations (Fig. 2.16). The process is guided by environmental and structural performance, design inputs, through which a bioclimatic technological system can be generated and implemented. The parametric digital model explored a series of differentiated geometries in terms of offset, torsion, and height of the vertical elements, starting from the extrusion of five ellipsoidal geometries.

The CFD performed simulation and the consequent genetic optimization allowed to verify the correspondence between morphology and execution, informing the next step of topological generation. Thanks to the 3D printing procedure, the digital model is directly transferred to the 3D printing machine through a design-to-production path, maximizing material consumption and saving production time. Compared to the traditional design and manufacturing process, the material represents a non-negligible parameter: the need to modulate fluid composition, extrusion pressure, layer's thickness, and correct speed of movements are design variables that should be considered not only in the final stage of fabrication but also in the project's early stages. The project foresees the incrementation of the structural performances using concrete and bamboo as reinforcement to be infilled in the empty segments of the

[8]CodesignLab (2016), "Ceramica Performativa", Available at: http://www.codesignlab.org/it/2014-02-09-22-48-47/90-homepage/244-ceramica-performatia.html (accessed 22 July 2018).

Triennale (2016). Available at: http://www.triennale.org/mostra/new-craft/ (accessed 9 August 2018).

vertical extrusion, as self-supporting complexes are able to assume different spatial and informed configurations. The analyzed case study concerns the experimentation of a ceramic component element and the process of digital computation aimed at generating informed geometries influenced by environmental and structural execution (project targets). The investigation aims at designing a passive cooling bioclimatic device that can also serve as a rainwater collector. The digital model is informed by the definition of a chain of variable and constraints; the latter gives rise to a catalog of design solutions where the shifting of a series of parameters describing the vertical elements corresponds to a sequence of responsive morphologies informed by the data utilized. The customization of the components allows defining specific patterns depending on the desired usage and on the performances required for applications on vertical opaque partitions and innovative multi-layer envelopes or in the context of redevelopment interventions. The 3D printing process constrains the dimensions of the components to the proportions of the machine, causing problems related to scalability and structural behavior of the technological system. In conclusion, the introduction of robotic fabrication protocol represents a possible improvement of the procedure described above (Fig. 2.17).

2.5 Conclusion: Technological Innovation for Informed Technological Systems and Traditional Materials

The computational and maximization process coupled with innovative digital manufacturing technologies opens new scenarios of investigation for the research on vast clay envelopes and the reintroduction of ancient Mediterranean construction techniques. The main alteration on the subject concerns the fabrication and construction protocols: from stratified and increasingly complex external partitions that, layer-by-layer, include the main structure, secondary structure, insulation, and cladding, we move to massive technological systems in which the formal generation is directly connected to the performance parameters that guide the design procedure. Technological innovation offers the possibility of experimenting with a renewed material sensitivity which can be used to explore new formal codes, and tectonics informed by quantitative performative criteria. The integration between modeling and robotic manufacturing adaptive technologies optimizes the component utilized only where necessary and based on the information of the design processes. With additive fabrication technology it is possible to assemble discretized elements and deposit material inspecting the kinematics of the robotic arms and the flexibility of parametric models. The methodology described above fosters the examination of maximized technological systems able to adapt to performative parameters. The analysis of the case studies has shown how, through new technologies, it is possible to reinterpret ancient construction techniques and traditional technological networks and utilize them to ensure the realization of customized architectures based on a specific material and a certain production process, energetically efficient, structurally performative, and

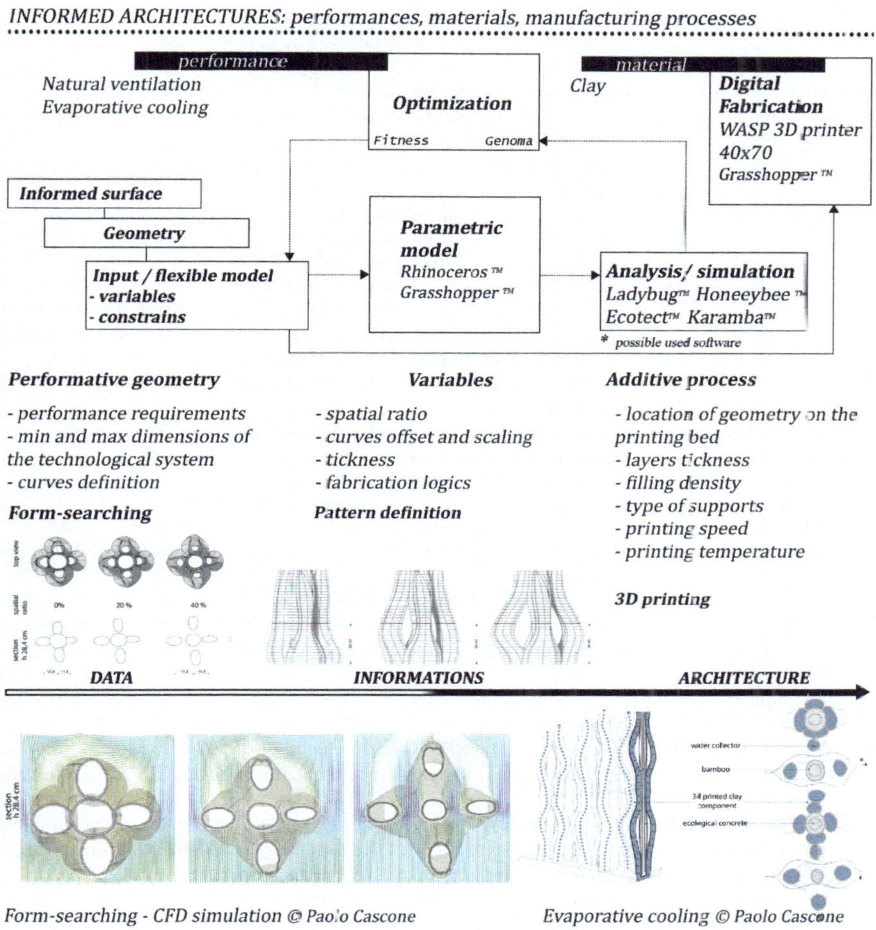

Fig. 2.17 *Ceramica Performativa*: performances, materials, manufacturing processes. A design methodology. © Angelo Figliola

morphologically responsive. Despite this, the application on enormous technological systems in clay in architectural practice is subject to the resolution of critical issues, such as

- Improvement of connection systems between technological units to ensure structural completion adequate to regulatory standards;
- Development of material mixtures with performance characteristics adapted to the standards and optimization of the production procedure;
- Assessment of performances related to material durability and static behavior during the post-production and installation phase.

References

1. Abufalia D (2013) Il grande mare. Storia del Mediterraneo, Mondadori, Milano
2. Battisti A (2014) Pensiero globale. Riqualificazione locale. Azione sociale. In: AA. VV., *LECTURES#2*, Rdesignpress. Roma, pp 180–197
3. Bottero M, Rossi G, Scudo G, Silvestrini G (1984) Architettura Solare. Tecnologie passive ed analisi costi benefici. CLUP, Milano
4. Braudel F (2002) Il Mediterraneo. Bompiani, Milano
5. Correia M, Dipasquale L, Mecca S (2014) Versus: heritage for tomorrow. Vernacular knowledge for sustainable architecture. Firenze University Press, Firenze
6. Crutzen P (2005) Benvenuti nell'Antropocene. Mondadori, Milano
7. Forlani MC (2004) La proposta di una nuova "Tecnica Antica". Tecnologie e manutenzione per i paesi in via di sviluppo. In: Missori A (eds) Tecnologia, progetto, manutenzione/ scritti sulla produzione edilizia in ricordo di G. Ferracuti, Franco Angeli, Milano
8. Hensel MU (2013) Performance-oriented architecture: rethinking architectural design and the built environment. Wiley, Chicester
9. Mecca S, Bati S, Forlani MC, Germanà ML (2011) Earth/Land. Edizioni ETS, Carrara
10. Rauch M, Kapfinger O (2001) Rammed earth: Terra Cruda. Birkhauser, Basel
11. Rykwert J (2008) La seduzione dei luoghi. Einaudi, Torino
12. Trombetta C (2002) L'attualità del pensiero di Hassan Fathy nella cultura tecnologica contemporanea. Il luogo, l'ambiente e la qualità dell'architettura. Rubbettino, Catanzato
13. Bechthold M, King N, Kane A (2015) Ceramic material system in architecture and interior design. Birkhauser Verlag Gmbh, Basel
14. Bechthold M, King J, Kane A, Niemasz J, Reinhart C (2011) Integrated environmental design and robotic fabrication workflow for ceramic shading systems. In: International symposium on automation and robotics in construction, proceedings of the 28th ISARC, Seoul, Korea, pp 70–75
15. Bechthold M (2013) Design robotics: new strategies for material system research. In: Peters B, Peters T (eds) Inside smart geometry. John Wiley & Sons, London, pp 254–267
16. Andreani S, Garcia del Castillo JL, Jyoti A, King N, Bechthold M (2012) Flowing matter: robotic fabrication of complex ceramic systems. In: Proceedings ISARC 2012. Eindhoven, The Netherlands
17. Andreani S, Bechthold M (2014) [R]evolving brick: geometry and performance innovation in ceramic building systems through design robotics. In: Gramazio F, Kohler M, Langenberg S (eds) FABRICATE: negotiating design & making. Gta Verlag, Zurich, pp 182–190
18. Tedeschi A, Wirz F (2014) AAD-Algorithms-aided design: parametric strategies using grasshopper. Le Penseur publisher, Brienza
19. Friedman J, Heamin K, Mesa O (2014) Woven clay. In: Gerber D, Huang A, Sanchez J (eds) Design agency: Projects of the 34th Annual Conference of the association for computer aided design in architecture, ACADIA. Riverside Architectural Press, pp 223–226
20. Friedman J, Kim H, Mesa O (2014) Experiments in additive clay deposition. In: Mcgee W, Ponce de Leon M (eds) Robotic fabrication in architecture, art and design 2014. Springer, New York, pp 261–272
21. Mostafavi S, Bier H, Bodea S, Anton AM (2015) Informed design to robotic production systems. In: Martens B, Wurzer G, Grasl T, Lorenz W, Schaffranek R (eds) Real time–proceedings of the 33rd eCAADe conference. Vienna University of Technology, Vienna, pp 287–29
22. Mostafavi S, Bier H (2016) Materially informed design to robotic production: a robotic 3D printing system for informed material deposition. In: Reinhardt D, Saunders P, Burry J (eds) Robotic fabrication in architecture, art and design 2016. Springer International Publishing, pp 339–349
23. Persiani S, Battisti A (2015) Componenti Innovativi in Argilla per l'Architettura Sostenibile: Elementi Massivi a Prestazioni migliorate. Edizioni Accademiche Italiane

Chapter 3
Informed Architecture and Wooden Structures. Overview of the Main European Research Paths

Abstract One of the main aspects investigated in the European research context on Informed Architecture is related to the use of digital innovations in wood structures construction of units and technological systems as well as architectural organisms at 1:1 scale. To analyze the different approaches, the contribution proposes a series of case studies and the results of two applied researches, the 1:1 scale pavilions *Fusta Ròbotica* and *Digital Urban Orchard*. The case studies are selected verifying the correspondence to the following parameters: the presence of a performance-based process through which explore informed architectures; the use of low-engineered and natural wood; the materialization of the digital model through new manufacturing processes, specifically robotic fabrication. The contribution allows gathering pros and cons in the three different investigative macro-areas: performance-based design, material culture, and fabrication process. This analytical inspection helps to create a clear research scenario around the topic of digital wood design in addition to the definition of an innovative pathway for future studies, as well as the assimilation of these novel concepts in the building construction sector.

Keywords Wood systems · Parametric design · Fabrication strategies · Parametric timber engineering · Informed timber structures · Optimization of wood architectures

3.1 Research Lines in the Field of Performative Wood

An important feature inquired in the European research context on Performative Architecture is connected to the utilization of original technologies in the field of wooden edifices for the production of technological systems in addition to architectural organisms at 1:1 scale. One of the oldest architectural materials which is widely utilized to this day is wood, this is because of its physical and mechanical properties, flexibility in structural employment which allows to use it indifferently to traction, compression, and flexion stresses. Compared to traditional wooden applications those related to digital computation, generative design, and robotic fabrication, open new horizons of research with different purposes, sharing the same investigative

A. Figliola and A. Battisti, *Post-industrial Robotics*,
https://doi.org/10.1007/978-981-15-5278-6_3

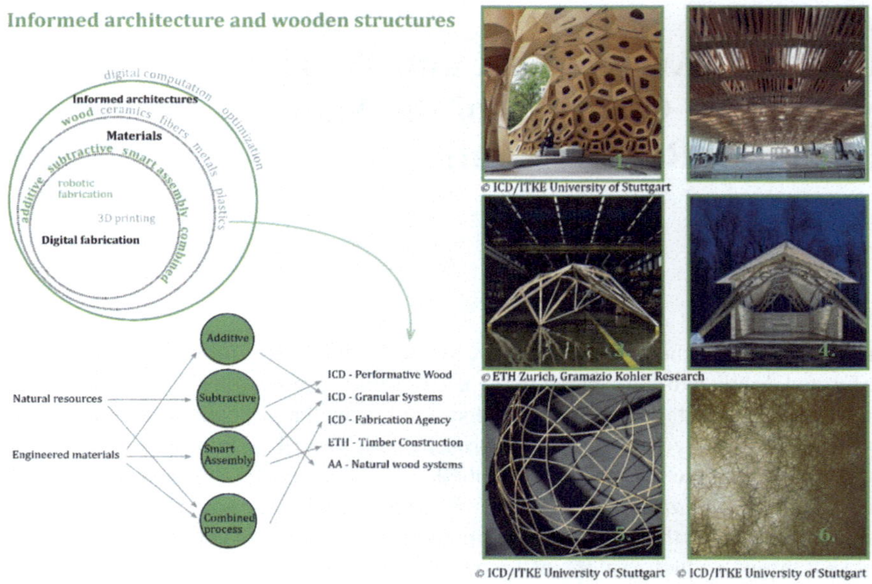

Fig. 3.1 Informed architectures and wood structures. Overview of the main research topics. © Angelo Figliola. Copyrighted material: © Gramazio Kohler Research and © ICD/ITKE University of Stuttgart

tools (Fig. 3.1). An overview of the theme is outlined by the texts Advancing Wood Architecture [20] and Advanced Timber Structures [24] that identify the main topics and the research centers involved to define the possible future research directions. The examinations conducted can be summarized in two distinct techniques: the first concerns the use of engineered wood elements to attempt to exploit its physical and mechanical properties; the second approach provides the utilization of wood as natural and low-engineered material for structural applications that benefit from the complex relationship between computational design and digital fabrication.

3.1.1 Wood Computation: Exploit Material Behavior as Design Agent

Regarding the first approach, one of the focus of the studies consists of the reviewing of mechanical assets of the material to transform them into design opportunities in the formal generation process. The computational method becomes the means by which to discover advantages offered by wood in relation to its complex and non-linear behavior. The physical and mechanical properties of this element, especially if used in the form of thin panels handled by a rolling procedure, such as heterogeneity, anisotropy, hygroscopic, and irregularity, can expand the range of design possibilities and stimulate the creative process. The studies on wood computation have been for some years part of the research conducted by Achim Menges at the University of Stuttgart, Institute for Computational Design (ICD) within the Performative Wood research line; by Michel Hensel at the Research Center for Architecture and Tectonics at the Oslo School of Architecture and Design. The investigations on the topic concern two performance aspects of the design that can benefit from the material qualities: the structural conduct and geometric-formal responsiveness [23]. Within this testing field different lines of research can be identified, and can be summarized as follows:

- Activation of the anisotropic properties of wood, to set distinct structural behaviors in relation to differentiation of stiffness in the direction of the fibers;
- Utilization of elastic assets of wood as a factor design, through bending procedures [18] to design not standard edifices;
- Exploitation of irregularity, through selective processes of digital fabrication to eliminate portions of inactive and inefficient material;
- Use of hygroscopic properties of wood to actuate morphological transformation procedures.

3.1.2 Wood Computation: Smart Assembly, Natural, and Low-Engineered Materials

A further aspect of research on the application of innovative technologies in wood constructions concerns the design and the prototyping of new spatial and structural configurations. This approach benefits from the digital design and robotic manufacturing with special attention to the smart assembly. The Institute of Technology in Architecture (ITA) of the Federal Polytechnic of Zurich (ETH) is one of the most active research centers on the topic of complex and optimized wood structures explored through the fabrication of a series of prototypes made in collaboration with construction industries [13]. The recent study developments on the subject focus on the possibility to extend the sensory capacity of the robot through a succession of sensors and a feedback loop system between the machine and the material system. A feedback loop means building relationships and designing a dynamic and non-linear complex capable of guaranteeing adaptability to the local scale. The machine becomes an extension of the craftsman who perceived and analyzed physical properties of a material through his own hands [7, 11]. The sensing ability of robots can be considered as the main novelty in the field of wood construction. This methodological technique constitutes a design response to the theory of soft system, or adaptable networks, and in continuous evolution whose dynamism is constantly fed by a flow of outside information. The capacity to act and react dynamically makes the system responsive to variations due to the interaction between the parties involved.

These aspects are investigated by the Institute for Computational Design in Stuttgart in a research path defined as Fabrication Agency [18]. Another field of examination conducted mainly at the Architectural Association (AA) School of Architecture in London and IAAC concerns the use of low-engineered and natural components present in great abundance in nature to create complex and optimized wood structures [20]. The project Wood Chip Barn from AA is a manifesto of this specific approach. The role of digital computing and the importance of data-driven strategy is evident: the scanning of the shrub and the consequent creation of a digital catalog based on dimensional and morphological variables represent the true innovation of the design process. Through the 3D scanning and reconstruction of the elements, it is possible to realize a complex surface starting from natural curvature and dimensional parameters of the trunks, avoiding expensive industrial bending procedures. The computational process allows to integrate with the project the data related to the natural material utilized, the structural conducts, and the variables of the building site. The relational models obtained by the generative process define the concept of informed morphology.

3.2 Performative Wood: Case Studies

To analyze the different techniques, the contribution proposes a series of case study and the results of two applied research, the 1:1 scale pavilions *Fusta Ròbotica* and *Digital Urban Orchard*. The case studies are selected verifying the correspondence to the following parameters:

- the presence of a performance-based procedure through which to explore informed architectures;
- the use of low-engineered and natural wood and the engineered wood;
- the materialization of the digital model through innovative manufacturing processes, specifically robotic fabrication.

All the projects are the results of a performance-based process [22] through which the generative process is informed by performative parameters (e.g., structural, environmental). This methodology enables the inquiry of novel formal codes utilizing different typologies of wood as construction material. The presence of innovative digital fabrication tools such as robots implies that the universities involved in the survey are the ones equipped with robotic manufacturing laboratories and who work to realize 1:1 scale projects.

Fig. 3.2 AA, *Wood Chip Barn*, Architectural Association: pavilion robotically fabricated and made by natural and low-engineered wood © Zachary Mollica

3.2.1 AA School of Architecture. Wood Chip Barn: Exploring the Potential of Natural Material

The case study *Wood Chip Barn*[1] (Fig. 3.2) constitutes the real transposition of a complex functional program through the use of a natural material, large wooden tree trunks, and an informed computational process. The project comes from the selection of 250 tree trunks grown in the Hooke Park forest that surrounds the homonymous Campus.[2] The tectonic and structural experimentation is informed by the 3D scan of the bifurcated wooden logs for the extraction of the geometric characteristics and organize the elements in the 3D space. From the scanning of the elements the following data are extracted: volumetry, central axes of the trunks with regard to the bifurcations present, and points of intersection between the axes themselves. The extricated information is utilized to create a digital catalog of solutions for the spatial organization of the logs in the formation of the main beams. The spatial organization of the trunks is maximized to achieve a structural system able to increase the structural stiffness and minimize shear stresses. The optimization process relies on the use of genetic algorithms through the definition of the fitness value, target, and variable parameters (genome). The arch structure, composed of different boards connected in the central points of the bifurcations, is anchored to the ground by foundations in concrete while the beams are pre-assembled by dividing the arc into two halves to be assembled in situ (Mollica and Self [21]. The robotic manufacturing procedure

[1]*AA Wood Chip Barn*. Students: Zachary Mollica; Swetha Vegesana; Sahil Shah; Vivian Yang; Mohaimeen Islam. Tutors: Martin Self, Emmanuel Vercruysse, Jack Draper, Charley Brentnall, Toby Burgess. With: Pradeep Devadass (Robotic Developer), Arup (Engineering).

[2]Hooke Park is a research campus in Dorset for experimental wood architectures of the Architectural Association of London, AA.

a

b

Fig. 3.3 **a** AA, *Wood Chip Barn*: complex joints made by robotic milling process **b** AA, *Wood Chip Barn*: complex joints made by milling process informed by data-driven design. © Zachary Mollica

that involved the utilization of a Kuka KR 150 robot with a milling spindle made it possible to create customized male-female joints between the different elements that make up the primary beams with 6 h of processing in the robot cell.

As a result of the use of the subtractive milling process, it was possible to realize male-female interlocking joints as well as joints for the installation of the secondary structure on which the wooden covering was installed. The proposed case study demonstrates in what way the new production technologies, together with a computational procedure informed by real-time data, it is possible to achieve optimized architectonic bodies that use non-engineered materials. The prototype is the result of a data-driven computational process able to utilize the components and structural performances as guiding parameters of the design, the acquisition of real-time data as feedback loop between the digital environment and the real one, and the genetic maximization to determine the best spatial configuration of structural elements. Through the 3D scanning process, the natural material, bifurcated wooden logs, is digitized to acquire the dimensional and geometric variables necessary to inform the digital model. The operative methodology has allowed to realize an optimized edifice able to

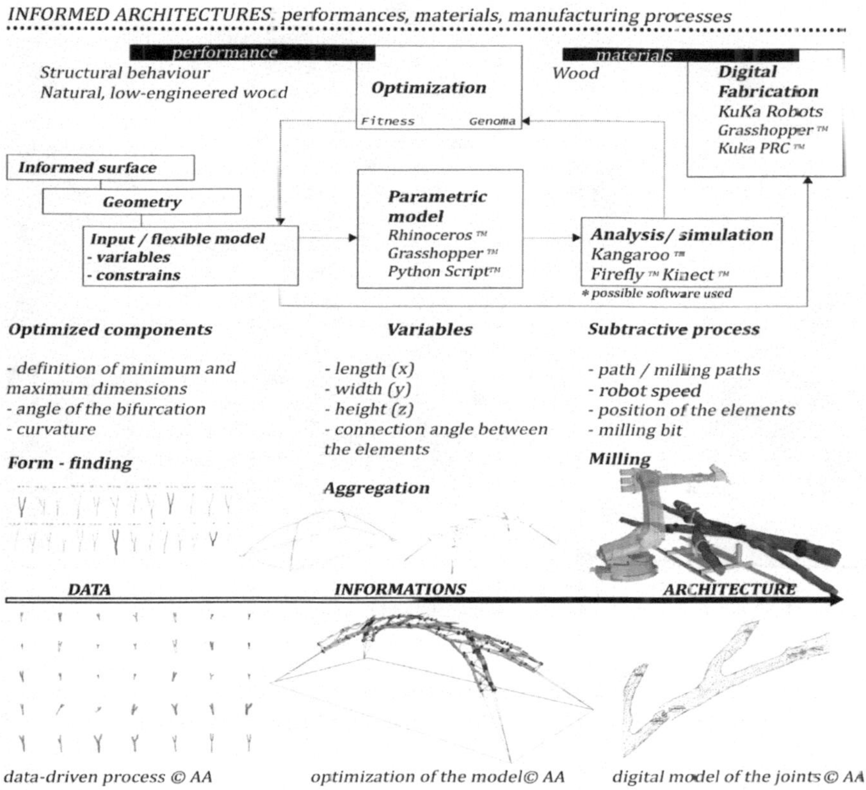

Fig. 3.4 AA, *Wood Chip Barn*: performances, materials, manufacturing processes. A design methodology. © Angelo Figliola

respond to performance parameters derived from the components, from the project site as well as from the structural conformation. In conclusion, the methodology applied to the Wood Chip Barn project has allowed to obtain an informed geometry whose materialization has been possible, thanks to a robotic manufacturing procedure (Fig. 3.4) through which to realize the customized connections (Fig. 3.3) in relation to the maximized spatial configuration. Non-engineered elements and a non-industrial production setting open up new tectonic speculations and formal codes for sustainable architectures.

Fig. 3.5 ICD/ITKE *Research Pavilion 2011*. © ICD/ITKE University of Stuttgart

3.2.2 ICD/ITKE Research Pavilion 2011: Discover Opportunities Offered by Wood Complex Behavior

The research pavilion 2011[3] is produced by the Institute for Computational Design and Construction (ICD) Stuttgart (Fig. 3.5) in collaboration with the Institute of Building Structures and Structural Design (ITKE) and other industry partners. The aim of the investigation is to propose a wooden structural system at 1:1 scale as a result of a computational and manufacturing processes informed by a series of performative parameters extracted from the analysis of a specific biological organism [14].

More specifically, the research involved the study of the skeleton's structure of the Echinoid (sea urchin) in relation to their composition and the rules of aggregation. The biological apparatus is examined to extract the physical and geometrical principles to be transferred to the structural-material system through the computational process. The analysis carried out on the biological organism has permitted the selection of fundamental geometric principles including the modularity of the system

[3]Institute for Computational Design—Prof. Achim Menges, Institute of Building Structures and Structural Design—Prof. Jan Knippers. Concept & System Development: Oliver David Krieg, Boyan Mihaylov. Detail Design & Fabrication & Construction: Peter Brachat, Benjamin Busch, Solmaz Fahimian, Christin Gegenheimer, Nicola Haberbosch, Elias Kästle, Oliver David Krieg, Yong Sung Kwon, Boyan Mihaylov, Hongmei Zhai. Scientific Development: Markus Gabler (project management), Riccardo La Magna (structural design), Steffen Reichert (detail design), Tobias Schwinn (project management), Frédéric Waimer (structural design).

that allows a high degree of adaptability and a performative structural system [16]. This organizational concept is conveyed to an optimized surface, obtained through a form-finding procedure integrated with structural scrutiny, from the tessellation of the same with hexagonal geometric components. The hexagonal elements are assembled with two different strategies: the plates that make up the single 3D technological unit are connected by structural glue while the assembly of the different units takes place due to the realization of 100,000 geometrically distinct joints and subsequent nailing process. The structure turns out to be very hierarchical and heterogeneous, since the size of the hexagonal cells are informed with regard to the curvature of the surface (anisotropic) as each hexagonal portion is oriented with attention to the mechanical-structural stress determined by the distribution of weights on the surface. The construction maximized in relation to structural performance parameters was achieved through a robotic manufacturing procedure and a mechanical processing of the subtractive, milling type, through which it was possible to guarantee the customization of the various connections, defined as finger joint. The proposed methodology has allowed the creation of an architectural body of considerable size using only plywood in panels with a thickness of 6.5 mm, creating 850 geometrically different components and 100,000 finger joint junctures between the individual elements. The research pavilion is the result of the transfer of biological principles in the project through digital computation and a robotic fabrication process (Fig. 3.6).

The proposed methodology is based on the information of the procedure with regard to data obtained from the analysis of the biological organism under investigation: scientific data, such as modularity and adaptability of the natural system, are transformed into geometric rules used to discretize an optimized surface through a form-finding process [19]. The result of this procedure is a heterogeneous responsive morphology utilized to realize an architectural structure in which the performance characteristics respond to mutable external inputs. The possibility of creating this structure through the assembly of different units made of extremely light wood panels, confirms the excellent structural performance that the architectural body can guarantee even if it requires a ground anchor to withstand the horizontal wind load. The major problems derive from the complex network of joints, the result of the heterogeneity and anisotropy of the material system, which makes the management of the production process complex compared to the generation of the manufacturing code and the speed of the fabrication procedure (Fig. 3.7). One of the possible implementations may derive from the simplification of the tessellation geometry and the connection system as well as the utilization of the entire volume of the hexagonal components.

Fig. 3.6 ICD/ITKE *Research Pavilion 2011*: tailored finger joints and assembled hexagonal cells. © ICD/ITKE University of Stuttgart

3.2.3 ETH, Gramazio Kohler Research. Topology Optimization and Robotic Fabrication of Advanced Timber Space-Frame Structures

The proposed case study [5] created by the ETH in 2015[4] on the field of additive robotic fabrication of complex building structures presents an optimized technological system (Fig. 3.8) with regard to the proper criteria of topological maximization commonly applied to complex buildings made with concrete or other fluid-dense

[4]The project was a research exchange between ETH Zürich and Aarhus School of Architecture in collaboration with the NCCR Digital Fabrication MAS Programme and Israel Institute of Technology, Haifa.

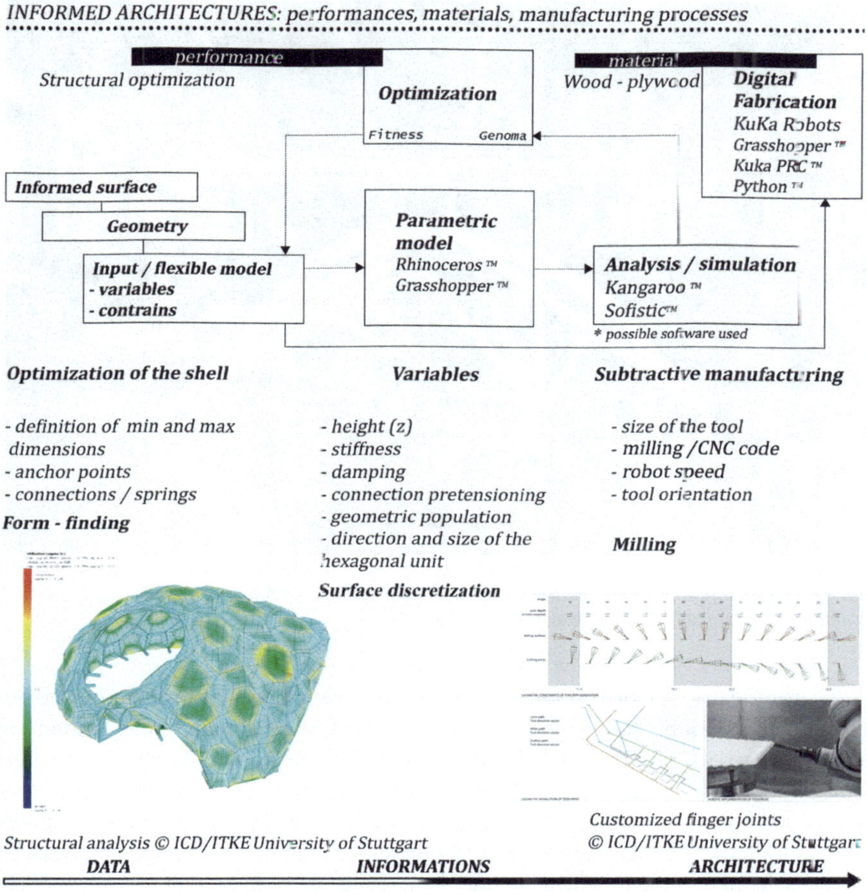

Fig. 3.7 ICD/ITKE University of Stuttgart, research pavilion 2011: performances, materials, manufacturing processes. A design methodology. © Angelo Figliola

materials. In this specific case, the optimization process is employed in a spatial structure composed of linear wooden structural elements described by the following characteristics: the spatial coordinates x, y, z which define the geometry at the macroscale, the structural support points, the localization of loads, and their size as well as different types of wooden beams.

The application of this design methodology, based on the use of genetic algorithms, allows to optimize the utilization of the material as well as to determine the best spatial configuration in relation to the employed loads and their distribution; the morphology obtained as a result of topological maximization will be able to guarantee the best structural performance with regard to the quantitative variables chosen as the objective of the process such as structural stiffness and stability (Fig. 3.9).

Fig. 3.8 ETH, *Topology Optimization and Robotic Fabrication of Advanced Timber Space-Frame Structures* © Gramazio Kohler Research, ETH Zurich

More specifically, for the proposed project the computational optimization procedure makes it possible to maximize the resistant section of each individual structural element in wood, eliminate the excess connections as well as establish the best configuration with regard to the load conditions. Unlike concrete structures, topological maximization applied to this type of discretized and non-continuous buildings leads to high complexity and to the customization of the bonds between the various components that make up the structure. As a result of the use of an industrial robot and a circular saw, it was possible to realize the customized spatial joints between the optimized wooden beams and to guarantee the manufacturing of all the distinct junctures. In addition to robotic fabrication, it was necessary to develop a building strategy able to eliminate the presence of support structures and assure stability in the continuation of operations. Through the digital model, a series of relationships are built to define assembly priorities in relation to the types of connections present. The prototype, composed of 34 wooden beams with dimension 5 m × 5 m × 5 m, has been subjected to structural checks, through load tests, in order to verify the reliability of topological optimization with regard to the digital model and prototype. The analyzed structural system explores the application of an informed computational strategy taking into consideration the structural data for the realization of a technological network composed of discretized and non-continuous linear elements. The verifications carried out following the realization of the prototype allowed to test the actual accuracy of the digital simulation defining, in fact, the process ready for any structural applications to the architectural scale. Despite this, some critical

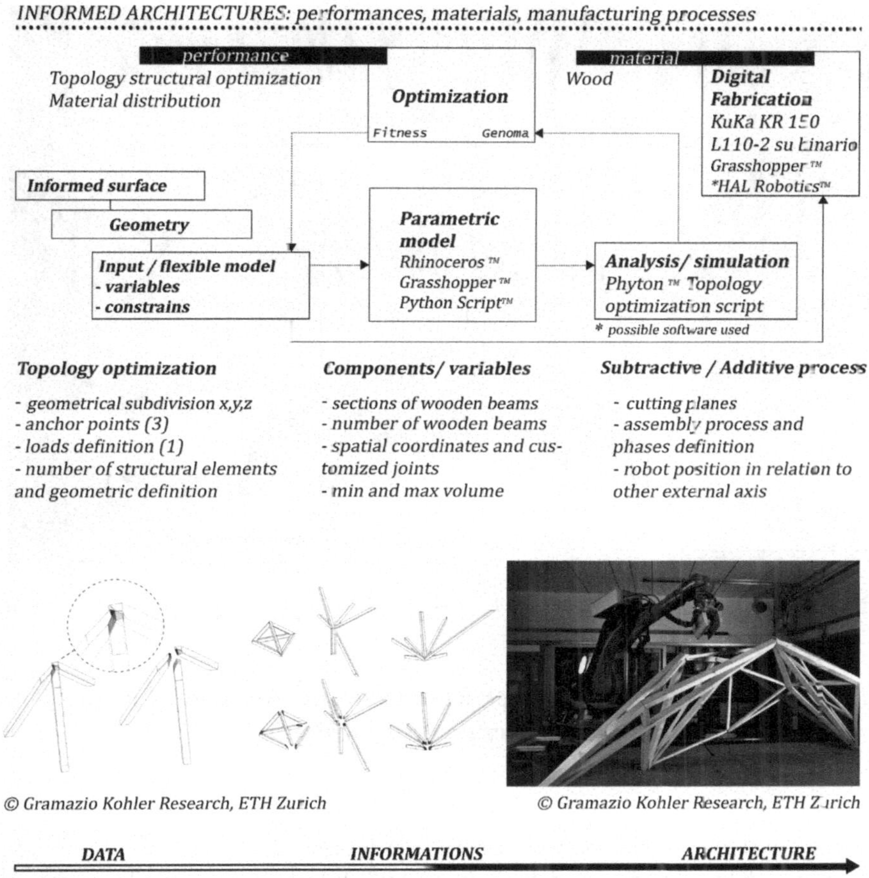

Fig. 3.9 ETH, *Topology Optimization, and Robotic Fabrication of Advanced Timber Space-Frame Structures*: performances, materials, manufacturing processes. A design methodology. © Angelo Figliola. Copyrighted material: © Gramazio Kohler Research

issues regarding the management of the computational procedure persist, increasing the geometric complexity (e.g., different load points, multiple loads, variable support points) and the manufacturing and assembly process. In the prototype realized, the dimensions of the components have facilitated man-machine collaborative assemblage maneuvers that would become difficult in the case of larger dimensional parameters. In this regard, the possible implementations concern the development of digital tools able to simplify the topological optimization procedure and the automation of the construction process.

Fig. 3.10 ETH Zurich, *Sequential Roof*, Arch_Tec_Lab of the Institute of Technology in Architecture (ITA). © Angelo Figliola

3.2.4 ETH, Gramazio Kohler Research. the Sequential Roof: Complex Timber Structures and Robotic Manufacturing

The examined case study[5] presents a responsive technological system (Fig. 3.10) for structural applications in 1:1 scale to be employed to free-form surfaces modeled by the Gramazio Kohler Research [12]. The 168 structurally independent reticular beams, composed of 23 layers assembled for a total thickness of 1,145 m with three types of wooden profiles, 50 × 115 mm, 50 × 140 mm, and 50 × 180 mm, and two support points, were utilized to create a coverage area of 2308 m^2 whose morphology was determined due to a structural and constructional maximization process [2]. The structural scrutiny allowed to optimize the morphology compared to the multiple load conditions while the density and the pattern of the structural elements were decided considering the presence of different technological subsystems (e.g., HVAC and MEP systems) and the need for inspection of the subsystems themselves. The

[5]Collaborators: Aleksandra Anna Apolinarska (project lead construction), Michael Knauss (project lead building project), Jaime de Miguel (project lead preliminary project), Selen Ercan, Olga Linardou.

Selected experts: Dr. Lüchinger + Meyer Bauingenieure AG, SJB. Kempter Fitze AG, Prof. Dr. Josef Schwartz (Chair of Structural Design, ETH Zurich), Prof. Dr. Andrea Frangi (Institute of Structural Engineering, ETH Zurich), Estia SA (EPF Lausanne), ROB Technologies AG. Selected contractors: Arch-Tec-Lab AG, ERNE AG Holzbau.

project workflow can be synthesized in four fundamental steps: geometrical modeling through a simple point and line model, structural analysis,[6] nail fitting, and a final phase of evaluation and eventually modification to ensure the feasibility of the project from design to production [4].

The investigations were conducted through digital models informed by the evaluations of scaled physical prototypes used to perform load analysis and structural test as well as to study the complex connections network (joints) and the relationship between the system and the external environment. More specifically, the permeability of natural light through the wooden structure has been ensured and evaluated by a series of testing utilizing a scale model and a sky-dome [3]. The workflow described above has allowed to inform the geometry with the data obtained both in digital and physical environments varying a series of variables such as the layering pattern of the structural beams, the cross section heights, the nail positioning, and the slat's end-cut area [4]. The customization of each beam requires an innovative and efficient constructive approach [17]. The setup of the construction process required a succession of tests with a multi-functional end effector able to position the wooden structural elements and automatically apply the polyurethane glue to then evaluate the tolerance of joints after operations of fixing. Due to a 1: 2 scale prototype it was possible to determine the criticality and potentiality of the procedure in relation to structural stability before scaling the whole process on an industrial level. The reviews carried out following the construction of the prototype have permitted to test the actual veracity of the simulation and the structural examination lead in the digital environment, defining the process ready for feasible applications to the architectural scale in the structural field, as well as for multi-layer wooden envelope. The design methodology (Fig. 3.11) and the resulting informed structural model was employed for the fabrication of the timber roof structure of the Arch_Tec_Lab building at the ETH campus Hönggerberg. The robotic manufacturing process for the realization of the roof was managed by ERNE AG Holzbau through Gantry Robot and an industrial endowment with the following characteristics [17]:

- 6 axes (2 parallel);
- dimensions: 48 m length, 5.60 m width, and 1.40 m height;
- 3 zones, each zone is 14 m long;
- maximum speed of 120 m/min (7 km/h);
- maximum load capacity of the gripper 250 kg;
- tools: circular saw, milling, closing, lifting;
- machine for wood, metal, and other materials;
- ability to determine the assembly sequence independently.

The *Sequential Roof* project explores the application of an informed design strategy with regard to structural data for the construction of a complex timber building to be employed in a 1:1 scale project respecting all the norm and code requirements and ensure the coexistence of different technological systems. Speaking

[6]The structural analysis is based on Finite Elements Analysis (FEA).

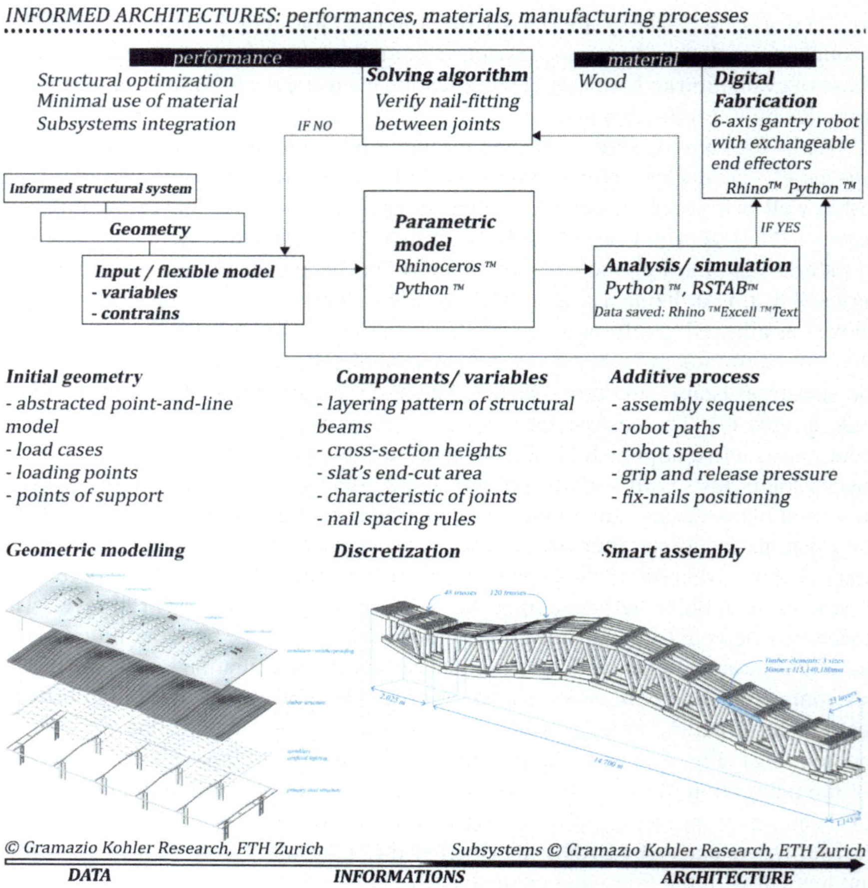

Fig. 3.11 ETH, *Sequential Roof*: performances, materials, manufacturing processes. A design methodology. © Angelo Figliola. Copyrighted material: © Gramazio Kohler Research, ETH Zurich

of which, the data management and the computational workflow were fundamental to produce fabrication data and manage the construction details based on codes.

3.2.5 ICD. Graded Light in Aggregate Structures

The proposed case study *Graded Light in Aggregate Structures*[7] [8, 9] (Fig. 3.12) develops a technological system based on the aggregation of small components,

[7]Desislava Angelova M.Sc., Thesis Advisers: Karola Dierichs, Thesis Supervisor: Prof. Achim Menges.

Fig. 3.12 *Graded Light in Aggregate Structures.* © ICD University of Stuttgart

technological units, able to self-determine the final morphology, and the resulting structural performance in a non-linear dynamic regime through an additive process [1, 15].

The small components described as granular materials[8] or particles, are able to create systems solely by contact forces between each technological unity. The clearly bio-inspired phenomenon, specified as an aggregate network, is studied and analyzed to define and extract the aggregation rules on the basis of which the morphology is determined and the structural performances are assured. In relation to the above, the construction of the prototype has introduced a new association between the technological system, material, and manufacturing process based on the concept of on-line programing. This methodology (Fig. 3.13) is founded on the alteration of the digital model by the real conditions of the physical environment under a dynamic regime due to the presence of sensors and a communication protocol between the digital and physical environment. The application of this design method allows the creation of responsive plane able to reply in real time to changes in the physical conditions based on the type of execution required, structural or energy preformative parameters. In particular, the proposed technological system is founded on a double curvature surface calculated and modeled starting from the comparison between the desired luminance conditions and the real values obtained thanks to the presence of a camera prepared for analysis. Through a real-time comparison of the two states expressed by bitmap images and color scales, the production process is adapted by varying the thickness of the surface and determining the pick and place path (displacement of elements for thickness variation) from the starting position toward the target point. The change in width and the arrangement of the aggregate components allows to obtain the preferred light gradient and therefore a geometry adept to respond to external inputs through morphological variations. An essential part of the procedure is the design of the end effector of the robot whose morphology is determined on the basis of the basic geometry that constitutes the aggregative system. The case study

[8]Examples of granular materials are sand, snow, rocks, etc.

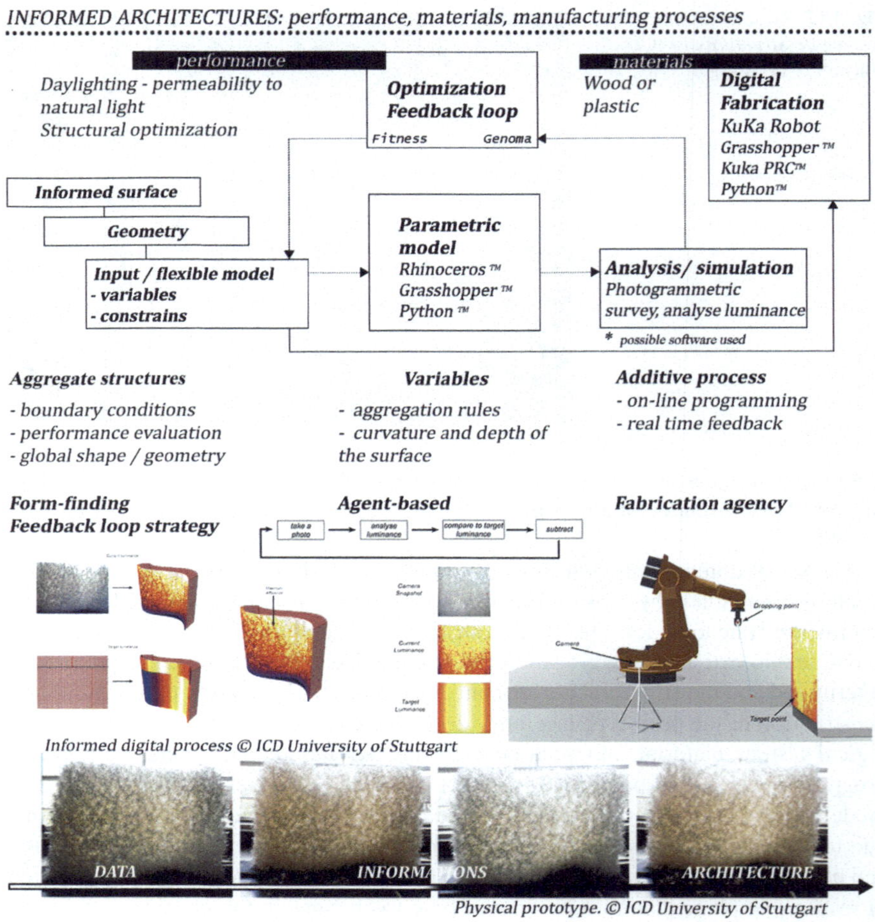

Fig. 3.13 ICD University of Stuttgart, *Aggregate Structures*: performances, materials, manufacturing processes. A design methodology. © Angelo Figliola

analyzed introduces the concept of on-line programing and real time transformation of the physical model in relation to the digital morphology and external inputs. In the proposed methodology, the concept of performance becomes a guide to the design process due to the utilization of data acquired from the physical conditions through the presence of sensors and specific communication protocols. The acquired data inform the design process in order to assure the correspondence between the requested executions and the conditions of the physical environment in which it operates. The information of the network is guaranteed by this continuous feedback loop and consequently the final geometry guarantees a perfect correlation between the digital model and the physical prototype. The procedure can be further implemented through the use of sensors placed directly on the end effector of the robot, increasing accuracy and speed of execution. Although the case study presented many critical issues, such as the management of the digital model and the iterations between digital and physical as well as the structural behavior of the system, it is an advanced example of geometric responsiveness where environmental performance, such as light permeability and solar radiation, is a guiding parameter for the morphology generation process. Starting from this thesis, ICD generates a sequence of research projects[9] to further explore the pros and cons of aggregate structures applied on architecture and construction. The key topics of the research were as follows:

- particles geometry, material, and fabrication rules;
- morphology generation according to structural performance and particle geometry;
- robotic manufacturing process and end effector development.

3.2.6 ICD. Robotic Softness: Behavioral Fabrication Process of a Woven Space

Robotic Softness[10] (Fig. 3.14) develops a network of relations (feedback loop) which regulates the formal generation with regard to execution parameters identified due to a customized and highly specialized production process [6]. The procedure commences from the study of the natural phenomenon of nesting, typical intertwined structures used as dens from birds, to analyze and understand the logic that manages the formation of such complex systems. Compared to the other proposed cases, *Robotic Softness* introduces a new concept that refers to soft systems characterized by a non-linear relationship based on continuous feedback both on a global and local scale. Just as in the volatile world, the formation of the interweaving that constitutes a nest is guided by environmental and structural performances, the digital and fabrication procedure that led to the prototype is informed on the global scale by evaluation variables such as light permeability, density, and thickness of the filaments, while at the local scale are the geometric parameters to inform the process with kinect sensor able, in real time, to return the data coming from the physical environment. As a result of the interaction between digital and physical worlds, the industrial robot can be considered as an *agent* suited to mediate with regard to the global scale and geometric variables at the local scale in a procedure that establishes only the boundary conditions and the general rules without defining the final morphology. For the development of the prototype, the generic industrial machine is equipped with a highly specialized end effector able to mediate the action taking into consideration the conditions of the local system in formation as well as the material used.

[9] ICD, Aggregate Pavilion 2015 and ICD Aggregate Wall 2017.

[10] Robotic Softness, Behavioural Fabrication Process of a Woven Space: Giulio Brugnaro, M.Sc. Thesis Advisers: E. Baharlou, L. Vasey. Thesis Supervisor: Prof. Achim Menges. Second Supervisor: Prof. Jan Knippers.

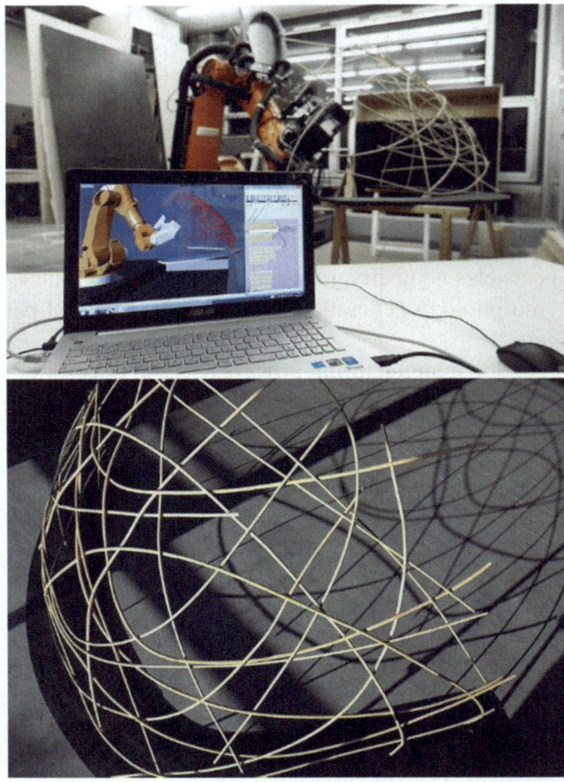

Fig. 3.14 ICD/ITKE, *Robotic Softness*. © Giulio Brugnaro, ICD/ITKE University of Stuttgart

The robotic network Agent-Based System (ABS) allows to adapt, setting boundary conditions and general rules, the virtual prototype to the materialization of the same in relation to data obtained in real time. The case study introduces an operational methodology in which the subject is in continuous transformation and its responsive behavior is regulated by real-time feedback with the physical environment. The environmental and structural performances constitute the essential variables of the data-driven strategy in such a way as to dispose the material, and consequently increase the density of the interweaving, where strictly necessary. This methodology allows to reinterpret an ancient method of processing materials, such as weaving, and an ancient component such as wood, through the fusion between digital model and physical conditions just as it happens in the formation of natural systems (Fig. 3.15). The proposed methodology and the geometry obtained presents a high degree of reactiveness with regard to external inputs, determined by the designer during the definition of the meta-project, with the optimization of the model in relation to the desired performances. The production of a customized end effector and the presence of a kinect sensor introduces the concept of cyber-physical making in which the machine

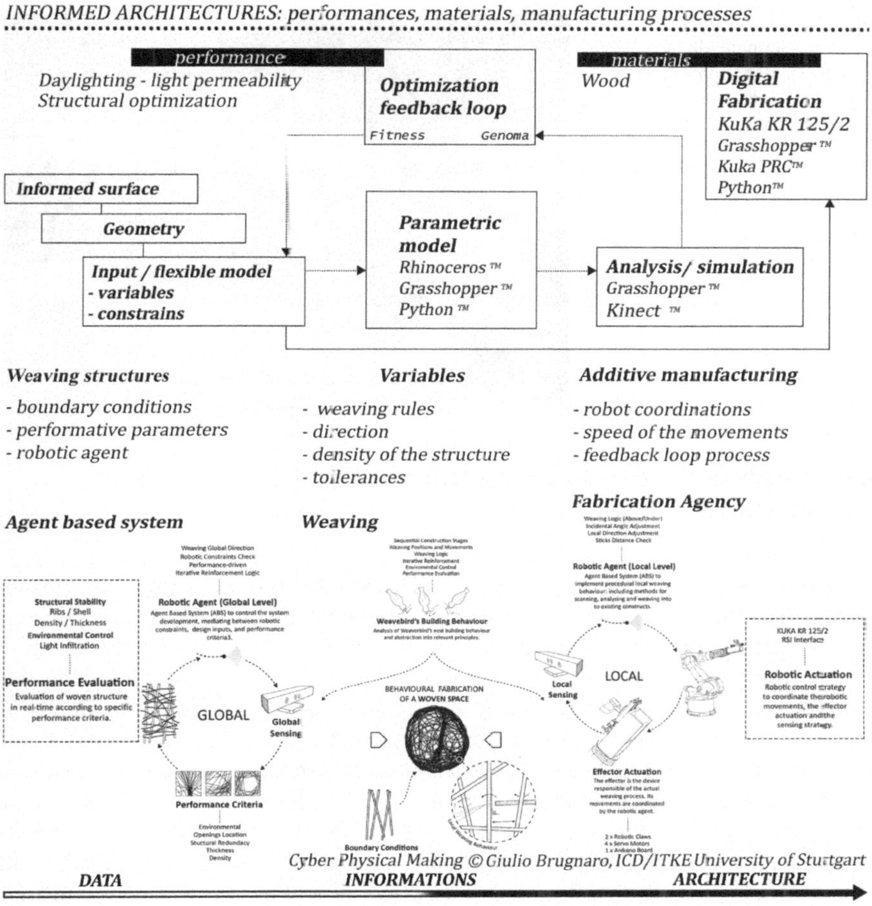

Fig. 3.15 ICD/ITKE University of Stuttgart, *Robotic Softness*: performances, materials, manufacturing processes. A design methodology. © Angelo Figliola

is not merely a performer of repetitive actions but interacts with the physical environment in a state of feedback loop between digital and physical environment. This kind of responsive system may be suitable for the realization of multi-layer casing elements capable of modulating natural light and solar radiation while guaranteeing energy savings and maximized utilization of the material. Among the critical issues it should be mentioned that the scalability of the system needs further testing and experimentation as well as the difficulty of managing the generative computational procedure and the robotic manufacturing process.

Fig. 3.16 IAAC, *Fusta Ròbotica*, Setmana de la fusta 2015, Barcelona. © Angelo Figliola, Open Thesis Fabrication

3.3 Applied Research: Fusta Robotica and Digital Urban Orchard

The prototypes, made in collaboration with industrial partners, represent the results of transdisciplinary experiments in which environmental, structural, and material performances inform the computational process and the robotics manufacturing.[11] *Fusta Robotica* is the outcome of a tectonic exploration deriving from non-engineered material, on the other hand *Digital Urban Orchard* is the formal expression of a complex functional program arising from the relationship among form (shape), function, and context. The practice of a new design paradigm based on the information of the process that sees environmental, structural, and material performances as a factor driving the entire design process. The execution criteria inform the computational procedure, subsequently materialized utilizing an anthropomorphic robot that is able to transpose informed digital models into physical reality, through non-industrial settings and using irregular and low-engineered elements [10].

3.3.1 Fusta Robòtica: Material-Informed Design

The *Fusta Robòtica*[12] pavilion (Fig. 3.16) is the first low environmental impact wooden structure built utilizing robotic manufacturing in Spain. It originated from

[11]The protorypes, Fusta Ròbotica and Digital Urban Orchard, was developed during the Open Thesis Fabrication 2015/2016 program at Institute for Advanced Architecture of Catalunya, IAAC. OTF team 2015–2016: Areti Markopoulou, Alexandre Dubor, Silvia Brandi, Djordje Stanojevic, Maria Kuptsova// Andrea Quartara, Angelo Figliola, Monish Siripurapu, Ji Won Jun, Josep Alcover Llubia, Yanna Haddad, Mohamad Mahdi Najafi, Fathimah Sujna Shakir, Nada Shalaby.

[12]Fusta Robòtica Pavilion is a project of IAAC, developed with the generous sponsor of Serradora Boix; in collaboration with Gremi de Fusters, Tallfusta, Incafust, Mecakim, Decustik.

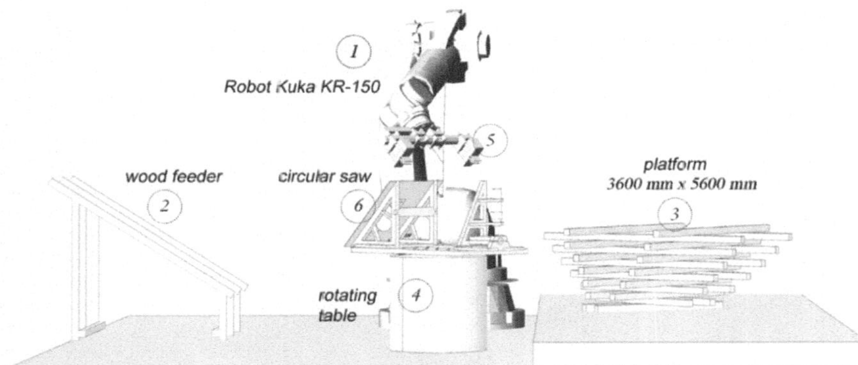

Fig. 3.17 IAAC, *Fusta Ròbotica*: setup of the robot cell and organization of the working space. © Angelo Figliola

a collaboration between the IAAC and Serradora Boix Srl as tectonic testing to be exposed at the Semana de la fusta 2015,[13] which has the intent to show the potential of the application of digital computation and robotics fabrication in the construction of wooden structures.

The objectives of the studies were represented by the promotion and enhancement of the Catalan wood (Mediterranean pine timber) by an exchange of knowledge between industries and research centers for the innovation of the production of the model in order to test new formal codes for a sustainable design. The material utilized, highly deformable and non-engineered, consists of small simple, and irregular wooden profiles and is employed for the production of industrial or biomass pallets. The square section profiles the industry has made available for the project have the following dimensions: 38 mm × 38 mm × 2000 mm. Another aspect to be integrated in the design process is the method of fabrication and the manufacturing area with their tools. The pavilion was built with a non-industrial setting, represented by a Kuka KR-150 industrial robot equipped with pneumatic gripper, a device for the storage of the wooden profiles and a circular saw arranged on a rotary table (Fig. 3.17), within the university digital fabrication laboratory.

The entire design procedure was informed by the mechanical properties of the material that were extracted through a series of analog tests necessary to the understanding of the component and structural system behavior. Among these, the excessive bending of the wooden profiles should be mentioned, due to the variation of the curvature following the drying process, and the lack of structural rigidity of the profiles as a result of the mechanical characteristics of the material. The manufacturing method adopted also contributes to inform the design procedure, with the minimum and maximum workable dimensions of profiles and components that can be aggregated depending on the work area, the characteristics of the robot, and

[13]Setmana de la Fusta is an annual event aimed to promote the use of wood Catalan that takes place in Barcelona.

Fig. 3.18 IAAC, *Fusta Ròbotica*: computational design process. © Angelo Figliola, Open Thesis Fabrication

their handling. In relation to this, the design process was informed using the material performances as a design input. It is possible to avoid structural problems due to the excessive bending of the wooden profiles using a redundant hyperstatic structure. Nailed joints allowed to maximize the resistant section of the elements, in correspondence of the structural nodes. The discretization of the shape in eight sections with a constant thickness optimized the working space and weight of the robot cell, avoiding collision problems and facilitating assembly. The pavilion, formed by about 1000 square section wooden profiles, is the result of the elaboration of a complex geometry (hyperboloid) in which the rotation of geometric continuous elements has permitted to obtain a dynamic spatial configuration, a manifesto of the potentials resulting from the use of material in the production of complex structures (Fig. 3.18). At the same time with the analog test development on the material system, the algorithm was developed to transpose the 3D solids of the digital model in simple geometric components, such as lines and planes, necessary for the definition of the various processing stages. Thanks to the direct connection between the parametric model and tool manufacturing, the different phases of the fabrication procedure have been defined. They can be summarized as follows:

- Taking off the wooden profiles from the storage device;
- Cutting of profiles to the corresponding size of the digital model;
- Profile deposit on the assembly platform.

Each stage of production included the construction of a half-arch to ease the operations of manual assembly and the transport to the installation site. At the end of the fabrication procedure 940 pieces of wood in 8 arches, divided into 16 parts were processed and put together in 35 h of production. The eight sections that make up the roof were assembled at the university laboratory and aggregated on the site of the installation.

Fig. 3.19 IAAC, *Digital Urban Orchard* 2016. © Angelo Figliola, Open Thesis Fabrication

3.3.2 Digital Urban Orchard: Form Follows Data Flow

The *Digital Urban Orchard* research project[14] (Fig. 3.19) involves the construction of a functional prototype to be implemented in urban public spaces within the self-sufficiency programme of the city of Barcelona, which stems from the relation among form, function, and application context for a new concept of space of socialization and food production. As second part of the project, the pavilion hosts a hydroponic cultivation system and an adaptive silicone skin able to ensure the indoor comfort conditions that are essential for the growth of plants. The need to design a stable yet lightweight structure and to ensure maximum solar gain for a proper growth of crops, at the same time, required multiple responsiveness capable of getting the proper compliance with the performance required by each of the single parameters listed above.

To integrate the functional, structural, and environmental-energy performance criteria, and inform the design process, the data-driven strategy was necessary to correctly set the genetic optimization by defining the genotype, the geometrical characteristics of the shape, and the phenotype or quantitative variables by which the genotype can be modified. The flexibility of the parametric model allowed to structure

[14]Digital Urban Orchard is a project of IAAC, developed with the generous sponsor of of Merefsa, supplying the silicone and with Windmill, in particular thanks to Josep Ramon Sole and Álvaro Romera for the structural consultancy.

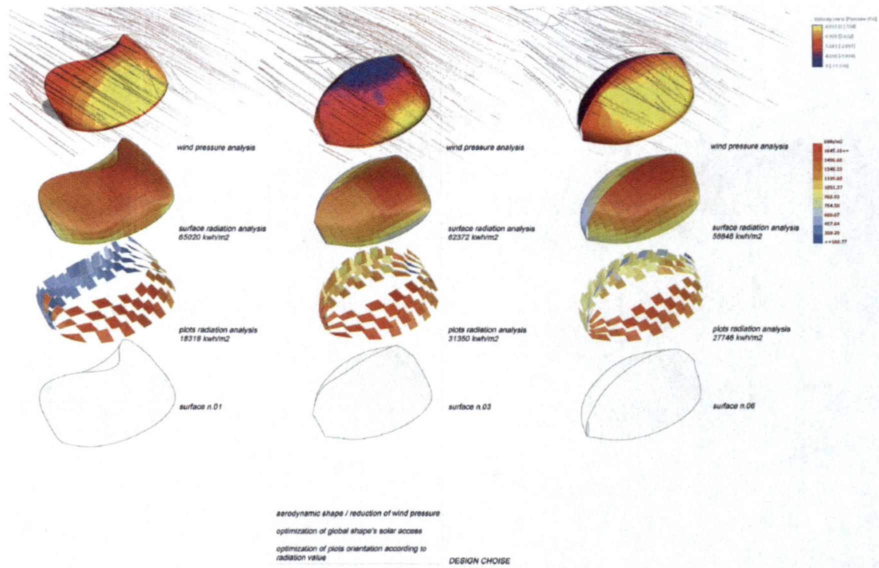

Fig. 3.20 IAAC, *Digital Urban Orchard*: digital catalog of informed architectures. © Angelo Figliola

the meta-project through the clarification of invariable parameters and genotype variable geometric data, which may vary within a range aptly delineated by the designer in relation to the values of the phenotype or rather quantitative variables of performance analysis. The final shape has been selected from a catalog of design solutions (Fig. 3.20), and is the result of genetic maximization and creative procedure which included the integration of different parameters:

- solar radiation on the surface of the orchard;
- solar radiation on the inclined planes where the plants are placed;
- wind pressure on the outer layer;
- minimum and maximum size of wood profiles that can be made with regard to the setting utilized;
- mechanical and physical properties of materials.

The process of genetic maximization was handled varying the geometric curves, two base ones on the x, y plane and a higher one, from which a surface is originated by the creation of a Loft and the inclination of planes that host the hydroponic system. The analysis of solar radiation on an annual basis, and the subsequent optimization, have made it possible to determine the overall shape and inclination of wooden shafts that host the hydroponic system. At the same time and due to a form searching procedure, the CFD exploration allowed to minimize the wind pressure on the outer surface of the pavilion in order to ensure structural balance. The adopted constructional principles are the same as those used in the *Fusta Robòtica* pavilion: the hyperstatic structural pattern, generated by the alternation of diagonals and elements able to ensure constructional rigidity, is a complex system which executes a structural function. It is designed as a support plan for the hydroponic system, as support for the silicone skin and as space-functional furniture. The density of the structural shapes responds to optimization logics for solar access into interior spaces and considers almost total transparency at the top of the pavilion. The final silhouette has been discretized through 6 types of segments, for 12 components total. Three fabricating strategies have been defined depending on the size of the sections and the work platform. They involve the construction of the entire segment or the joining of two/three parts of the final section with a total of 30 assembled parts. To maximize the resistant section, 2,524 nails were utilized in nailed joints with a collaborative process between manufacturing robotic and manual finishing. The structural analysis, conducted in cooperation with the engineering firm *Windmill* project, has enabled the validation of the constructional choice made despite showing a high displacement due to the horizontal pressure of the wind in extreme conditions as set forth by the legislation. In the *Fusta Robòtica* pavilion the fabrication process was effectuated at all stages in order to reduce material consumption and expand the range of achievable geometry. Implementations concerned the customization of end effector, pneumatic gripper, and tools used for the production such as the circular saw and the device for the storage of wooden profiles (Fig. 3.21).

The custom-built circular saw has allowed to create spatial cuttings (Fig. 3.22) in three dimensions. Thanks to a new wood provider, the lengths of the profiles were diversified in order to reduce waste material. Realized with 1681 profiles, the pavilion is the result of 52 h of robotic and 24 h of manual assembly coming from the information of the procedure and the optimization of the performance completed

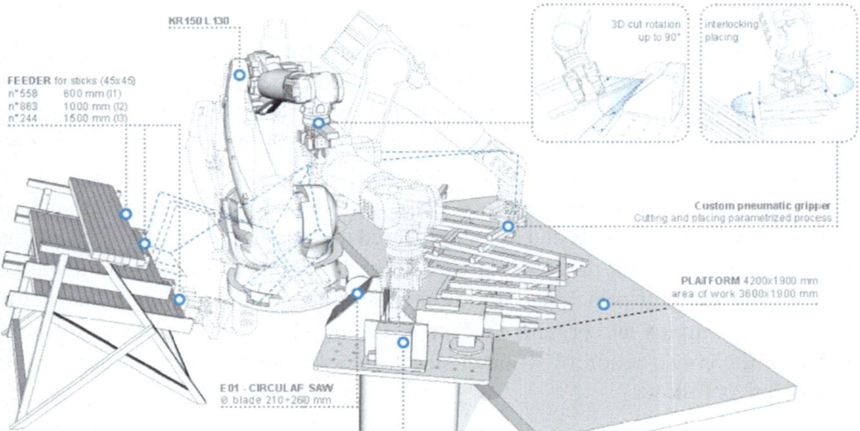

Fig. 3.21 IAAC, *Digital Urban Orchard*: setup of the robot cell and organization of the working space. © Angelo Figliola, Open Thesis Fabrication

Fig. 3.22 IAAC, *Digital Urban Orchard*: complex assembly thanks to the spatial cutting process. © Angelo Figliola, Open Thesis Fabrication

in a production process that can control the complexity and transform it into design opportunities while ensuring rapid execution, automation, and only 2% of material waste. Finally, during construction, the silicone wrap manufacture benefited from the collaboration with the silicone industry (Merefsa S.R.L.) that provided the necessary substances and laser cutting tools for the proper production of components.

3.4 Conclusion: Informed Wood Systems

The survey conducted on informed wood systems has demonstrated the possibility of concrete applications in architectural practice specially in the field of wood structural networks. The research paths on informed wooden buildings benefit from a consolidated know-how for what concerns the properties of the material and the related manufacture technologies by the whole company involved in the design process. The collaboration between academia and construction industries leads to introduce innovative technologies like digital computation and robotic fabrication to fully explore new formal codes related to environmental sustainability able to positively act on performative criteria such as the carbon footprint and the embedded energy of the procedures. The experiments examined employed engineered components such as laminated and profiled panels and low-engineered materials while digital exploration concerned computational processes of structural and multi-objective maximization as well as the study of principles of biomimesis to be abstracted and transferred in the design of complex and organic buildings. The digital manufacturing procedures entailed in the construction of wooden structural systems are different and can be summarized as follows:

- Reductive processes, for the realization of differentiated joints derived from complex optimized spatial configurations;

- Additive procedures,[15] for the assemblage of elements by means of composite and informed spatial sequences;
- Combined, subtractive, and supplementing processes, in which the construction succession involves machining the components before their assembly.

3.4.1 Performative Wood and Structural Systems: Toward an Application in Architecture

Some of the wood structural systems analyzed tested the integration between different technological networks such as the main structure, secondary structure, insulation, finishing, and technological systems. The capability to integrate complex technological networks through a digital workflow demonstrates the applicability of the methodology and validates the processes in relation to the management of tolerances derived from the use of innovative tools in the fabrication. Projects such as *Sequential Roof* of the ETH Zurich, roof structure of the Arch_Tec_Lab of the Institute of Technology in Architecture (ITA), as well as the entire production of the research path on additive procedures for the construction of compound wooden edifices of the Swiss university, *Wood Chip Barn* of the Architectural Association of London (AA), and the 2011 Research Pavilion of the ICD, basis for the construction of the *Landesgartenschau Exhibition Hall* of 2014, have demonstrated the efficiency of the computational process in the exploration of different design solutions and the concrete applicability of technological networks in correlation to complex systems that characterize an architectural organism. The described experiences can be considered as proof of concept for what concerns:

- The scalability of procedures and the integration of main and secondary technological systems;
- The potentiality of the morphogenetic process and digital computation in the transfer of biological principles in the virtual space, and subsequently physical;
- The advantages derived from genetic and heuristic optimization procedures in the design of performance structural systems;
- The potentiality of the customization of structural joints for the construction of optimized structures;
- The management of tolerances resulting from the digital manufacturing processes used.

A further check was carried out through the fabrication of two research pavilions, *Digital Urban Orchard* and *Fusta Robòtica*, made at the IAAC in Barcelona. The design of the two pavilions highlighted a further aspect regarding the applicability of structural wooden systems in architectural practice: the control of formal generation through the parametric process and the possibility of customization offered by the robotic fabrication, allows to overcome, or better expand, the concept of structure to

[15]The assembly of the components can be assimilated to other additive processes.

that of an integrated system that manages to aggregate the primary and secondary construction, as well as systems and furnishings through specific morphological configurations. The studies on the issue present a high level of employment also for the direct involvement of the industries in the sector stimulated by the sharing of know-how and the possibilities offered by the combination of a traditional material such as wood and digital processes of performance simulation and digital manufacturing. Compared to the projects mentioned above, other design experiences have a more experimental purpose, in some cases mere speculations, which corresponds to a low level of applicability in architectural practice, mainly in relation to the scalability of procedures. Nevertheless, starting from the analysis of the projects, it is possible to extract theoretical concepts and future applications that can be transferred to the design processes: one of these is the investigation of the soft systems, which contrasts with the hard systems discussed above, based on the adaptive relationship of feedback loop between performance parameters and formal generation. The studies that combined responsive morphologies with a robot directly on the construction site clearly showed the potential application in the realization of structural technological networks, or vertical ones, performance closures at the architecture scale.

Acknowledgments Fusta Robòtica Pavilion is a project of IAAC, developed with the generous sponsor of Serradora Boix; in collaboration with Gremi de Fusters, Tallfusta, Incafust, Mecakim, Decustik.

Digital Urban Orchard is a project of IAAC, developed with the generous sponsor of Merefsa, supplying the silicone and with Windmill, in particular thanks to Josep Ramon Sole and Álvaro Romera for the structural consultancy.

OTF team 2015–2016: Areti Markopoulou, Alexandre Dubor, Silvia Brandi, Djordje Stanojevic, Maria Kuptsova// Andrea Quartara, Angelo Figliola, Monish Siripurapu, Ji Won Jun, Josep Alcover Llubia, Yanna Haddad, Mohamad Mahdi Najafi, Fathimah Sujna Shakir, Nada Shalaby.

ICD/ITKE University of Stuttgart Projects Credits Robotic Softness. Behavioural Fabrication Process of a Woven Space: Giulio Brugnaro, M.Sc.; Thesis Advisers: E. Baharlou, L. Vasey; Thesis Supervisor: Prof. Achim Menges; Second Supervisor: Prof. Jan Knippers.

ICD/ITKE Research Pavilion 2011: Institute for Computational Design—Prof. Achim Menges; Institute of Building Structures and Structural Design—Prof. Jan Knippers; Concept & System Development: Oliver David Krieg, Boyan Mihaylov; Detail Design & Fabrication & Construction: Peter Brachat, Benjamin Busch, Solmaz Fahimian, Christin Gegenheimer, Nicola Haberbosch, Elias Kästle, Oliver David Krieg, Yong Sung Kwon, Boyan Mihaylov, Hongmei Zhai; Scientific Development: Markus Gabler (project management), Riccardo La Magna (structural design), Steffen Reichert (detail design), Tobias Schwinn (project management), Frédéric Waimer (structural design); Funding: KUKA Roboter GmbH, OCHS GmbH, Landesbetrieb Forst Baden-Württemberg (ForstBW), KST2 Systemtechnik GmbH, Leitz GmbH & Co. KG, Stiftungen LBBW, Müllerblaustein Holzbau GmbH, Herrmann Rothfuss Bauunternehmung GmbH & Co., Ullrich & Schön GmbH, Holzhandlung Wider GmbH & Co. KG.

Graded Light. Modulating the daylight in aggregate structures using online controlled robotic processes: Desislava Angelova M.Sc.; Thesis Advisers: Karola Dierichs; Thesis Supervisor: Prof. Achim Menges; First Examiner: Prof. Achim Menges; Second Examiner: Prof. Jan Knippers.

References

1. Angelova D, Dierichs K, Menges A (2015) Graded light in aggregate structures—modelling the daylight in designed granular systems using online controlled robotic processes. In: Martens B, Wurzer G, Grasl T, Lorenz W, Schaffranek R (eds) Real time—proceedings of the 33rd eCAADe conference, vol 2. Vienna University of Technology, Vienna, pp 399–406
2. Apolinarska AA, Bärtschi R, Furrer R, Gramazio F, Kohler M (2016) Mastering the "Sequential Roof"—Computational methods for integrating design, structural analysis, and robotic fabrication. In: Adriaenssens S, Gramazio F, Kohler M, Menges A, Pauly M (eds) Advances in architectural geometry 2016, vdf Hochschulverlag AG an der ETH Zürich, pp 240–258
3. Apolinarska AA, Knauss M, Gramazio F, Kohler M (2017) The sequential roof. In: Menges A, Schwinn T, Krieg OD (eds) Advancing wood architecture. Routledge, Oxon
4. Apolinarska AA (2018) Complex timber structures from simple elements. Computational design of novel bar structures for robotic fabrication and assembly. PhD thesis, ETH Zurich, Zurich, Swiss
5. Asbjørn S, Amir O, Eversmann P, Piskorec L, Stan F, Gramazio F, Kohler M (2016) Topology optimization and robotic fabrication of advanced timber space-frame structures. In: Reinhardt D, Saunders R, Burry J (eds) Robotic fabrication in architecture, art and design 2016. Springer International Publishing, pp 190–203
6. Brugnaro G, Baharlou E, Vasey L, Menges A (2016) Robotic softness. In: Velikov K, Sean Ahlquist S, Matias Del Campo M (eds) Proceedings of the 36th annual conference of the association for computer aided design in architecture, Ann Arbor, MI
7. De Landa M (2015) The new materiality. In: Menges A (ed) Material synthesis: fusing the physical and the digital. Wiley, London, pp 16–21
8. Dierichs K, Angelova D, Menges A (2015) Modelling Aggregate Behaviour. In: Thomsen R, Tamke M, Gengnagel, M, Faircloth C, Scheurer F (eds) Modelling behaviour—Proceedings of the design modelling symposium Copenhagen 2015. Springer, Cham, Heidelberg, New York, Dordrecht, London, pp 5–15
9. Dierichs K, Menges A (2015) Granular morphologies—Programming material behaviour with designed aggregates. Arch Des 85(5):86–91
10. Figliola A (2016) Post-industrial robotics: esplorazione di architetture informate nell'era post—digitale Techne. https://doi.org/10.13128/techne-xxxxx
11. Gramazio F, Kohler M (2008) Digital materiality in architecture. Lars Muller Publisher
12. Gramazio F, Kohler M (2014) Made by robots: challenging architecture at a larger scale. Wiley, London
13. Gramazio F, KohlerM Willmann J (2014) The robotic touch: how robots change architecture. Park Books, London
14. Jodido P (2013) ICD/ITKE research pavilion 2011. In: Jodido P (ed) Wood architecture now! Taschen GmbH, Köln, pp 194–199
15. ICD 2014 ITECH M.Sc 2014 Graded light. http://icd.uni-stuttgart.de/?p=11783. Accessed 22 July 2016
16. Knippers J, Menges A, Gabler M, La Magna R, Waimer F, Reichert S, Schwinn T (2012) Fromnature to fabrication: biomimetic design principles for the production of complex spatial structures. In: Hesselgren L, Sharma S, Wallner J, Baldassini N, Bompas P, Raynaud J (eds) Advances in architectural geometry 2012. Springer Wien, New York, pp 107–122
17. Krammer M (2016) Individual serialism through the use of robotics in the production of large-scale building components. In: Reinhardt D, Saunders P, Burry J (eds) Robotic fabrication in architecture, art and design 2016. Springer International Publishing, pp 461–467
18. La Magna R, Schleicher S (2016) Bending-active plates: form and structures. In: Adriaenssens S, Gramazio F, Kohler M, Menges M, Pauly M (eds) Advances in architectural geometry 2016. Vdf Hochschulverlag AG an der ETH Zürich, pp 170–188
19. Menges A (2013) Morphospaces of robotic fabrication. In: Brell-Çokcan S, Braumann J (eds) Rob/Arch 2012: robotic fabrication in architecture, art and design. Springer, New York, pp 29–47

20. Menges A, Schwinn T, Krieg OD (2016) Advancing wood architecture: a computational approach. Routledge, New York
21. Mollica Z, Self M (2016) Tree fork truss: geometric strategies for exploiting inherentmaterial form. In: Adriaenssens S, Gramazio F, Kohler M, Menges M, Pauly M (eds) Advances in architectural geometry 2016. Vdf Hochschulverlag AG an der ETH Zürich, pp 138–154
22. Oxman R (2009) Performative design: a performance-based model of digital architectural design. Environ Plann B Plann Des J36(6):1026–1037
23. Persiani S, Battisti A, Wolf T (2015) Recurring moving patterns in nature for a biomimetical optimization of autoreactive systems. In: Proceedings of the international association for shell and spatial structures (IASS) symposium 2015, Amsterdam
24. Weynand Y (2016) Advanced timber structures. Birkhauser, Basilea

Chapter 4
Informed Architecture and Plastic Materials. Overview of the Main European Research Paths

Abstract A new branch of investigation in the European research context on Informed Architecture is related to the use of digital innovations and plastic elements to design and fabricate units and technological networks as well as architectural organisms in a 1:1 scale. In order to analyze this path, the chapter proposes a series of case studies in which informed morphologies are realized by advanced digital production processes utilizing plastic material systems and their derivates. The selected research outlines the topics of examination and their field of application in the AEC sector. The primary focus is placed on indoor and outdoor performative structures and optimized technological devices able to improve the space comfort perception through intricate geometric configurations. The contribution allows gathering pros and cons in the three different investigative macroareas: performance-based design, material culture, and fabrication process. This analytical examination helps to create a clear research scenario around the topic as well as the definition of an original pathway for future studies, foreseeing the application of these innovative concepts in the building construction sector.

Keywords Plastic systems · Parametric design · Fabrication strategies · Additive manufacturing · Informed structures · Smart assembly

4.1 Plastics Materials and Digital Innovations

The use of plastic materials, and their derivatives, in the main European research context on the themes of informed architectures and digital manufacturing, stands out for the purposes acquired from their employment in the design process. According to the projects analyzed, plastic polymers such as the acrylonitrile-butadiene-styrene (ABS) and the polyatomic acid (PLA) are widely utilized in additive procedures, expanded polystyrene (EPS) in subtractive processes, and polymethylmethacrylate in formative procedures. They can be traced back to experimental needs of particular research paths as well as to the economic efficiency obtained from their use, which allows to carry out tests on 1:1 scale prototypes, even if the material is not suitable for possible applications in architectural practice. Based on the common

A. Figliola and A. Battisti, *Post-industrial Robotics*,
https://doi.org/10.1007/978-981-15-5278-6_4

utilization of plastic materials in architecture as insulation for vertical and horizontal partitions, façades and roofs, like ETICS, window frames, and functional components [5], recent technological innovations in the production field allow to expand the employment range of plastics elements and ensure a high geometric freedom compared to traditional manufacturing processes. Referring to the three large families implemented in the AEC sector, polyvinylchloride (PVC), polyethylene (PE), and polystyrene (PS), the procedures studied involve the use of new materials, at least for what concerns their utilization in the production of building components among which it is possible to mention thermoplastic polymers and bioplastics. Plastic polymers, which are plastic materials of high molecular weight, represent a new resource for the realization of technological units. The most commonly utilized type of polymers is that of thermoplastic, capable of acquiring malleability through the action of a heat source. The latter makes the material viscous and able to assume a great variety of morphological conformations and then return rigid; hence acquiring structural rigidity as the temperature decreases. This feature makes them theoretically reusable several/infinite times ensuring the recyclability of the material. The use of thermoplastic polymers in of technological components' fabrication process benefits from: the properties of the substance as described above; the opportunity of customizing the tools employed for the production path; the high degree of freedom due to the use of robotic arms [6]; the complexity of shapes which are possible to design and build. The design of a specific digital workflow and a highly custom-made end effector for the robotic arm is necessary to realize a fabrication process capable of exploiting the robotic manufacturing technology and the properties of the material. In addition to 3D printing, other digital fabrication procedures also profit from the generic nature of robots and the possibility of creating special instruments for manufacturing. Processes such as thermoforming, hot wire cutting, or the assembly of plastic components, benefit from the direct connection between the digital model and physical space as well as the prototyping of specific end effectors to carry out the production procedure.

4.1.1 Research Paths on Digital Fabrication with Plastic Polymers

Some of these aspects are addressed in the Mesh Mold research project of the ETH Zurich, which led to the definition of an extruded spatial wireframe structure able to function as a reinforcing element of the concrete and, at the same time, as formwork. The methodology proposed by Gramazio Kohler Research is aimed at studying a new technique of concrete casting and set up the reinforcement by replacing commonly used steel bars and wood panels with a unique spatial construction achieved by a robotic additive process. For the correct development of the research, it was necessary to design and realize a special end effector able to extrude with precision the thermoplastic polymer applied to the structure without utilizing expensive supports.

The extrusion device was prototyped and patented in cooperation with the Sika Technology AG industry. Although the plastic polymers used do not have physical and mechanical properties of structural elements, in the testing phase they allow defining the setting for the digital and robotic manufacturing process. Experimenting with innovative production methods adds to the possibilities offered by assembly procedures that benefit from the positioning accuracy and repeatability of execution offered by industrial robots [7, 8]. In further consideration, this technology can be considered as an additive fabrication process, based on the selective positioning of various materials to achieve a specific spatial configuration which meets the criteria of optimization of performance parameters and fitness criteria of the design procedure. Given the general nature of the machine and the possibility of customizing the end effector of robotic arms, this process can be employed to operate a series of actions, as assemblage and fixing or assembly and welding, simultaneously. The properties of plastics in combination with additive and subtractive manufacturing methods opens up to a succession of experiments related to the data-driven process, in which the distribution of the material is determined by the simulation and maximization of the performance parameters chosen as project objectives. Through this procedure the material is deposited or removed depending on structural or energy-environmental executions that guide the design process, generating a change of paradigm compared to the traditional concept of performance-based architecture. The manufacturing methods used in these procedures are various, from milling to 3D printing, and do not respond to specific research lines. As previously mentioned, the physical and mechanical plastic material properties allow a relatively simple application (especially thermoplastic polymers). Most of them are being tested in the research carried out by MIT, TU Delf-Hyperbody, Sydney University, and Tongji University that propose visually static responsive morphologies able to control the distribution of the material based on performative criteria. Furthermore, an interesting research path is the one carried out by the Design Informatics Department of the University of Delft that relates informed morphologies, plastic elements, and additive processes. Additionally, the innovation of the procedure can also be identified in the democratization of the manufacturing process through the diffusion of small 3D printers utilized directly from home. The designer sets up a digital procedure that customizes the morphology according to specific dimensional data, like performance parameters, delocalizing the production from industrial factories to the user's homes.

4.2 Case Studies

To analyze the different approaches the chapter proposes a series of case studies (Fig. 4.1) selected based on their correspondence to the following parameters:

- the presence of a performance-based process to explore informed architectures;
- the use of thermoplastics polymers and common plastics derivatives;
- the materialization of the digital model through innovative manufacturing procedures, specifically robotic fabrication.

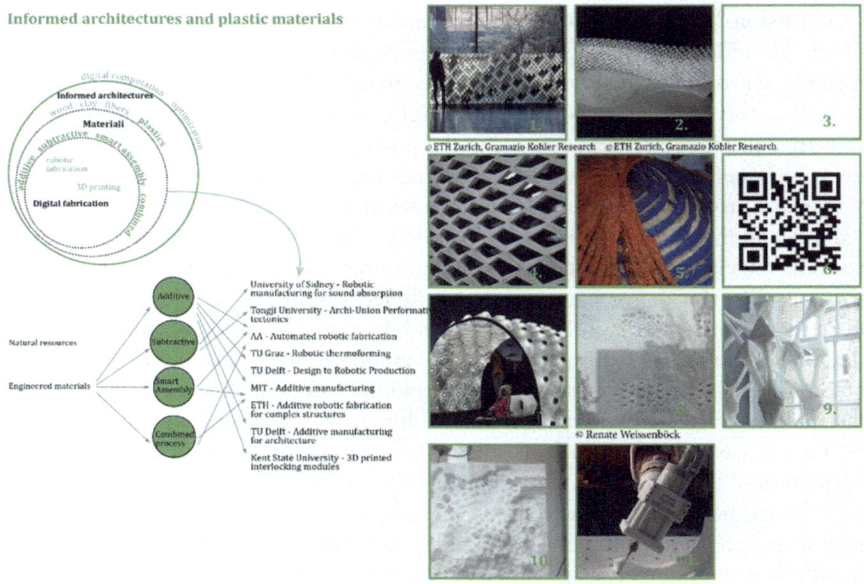

Fig. 4.1 Informed architectures and plastic materials. Overview of the main research topics and case studies. © Angelo Figliola. Copyrighted materials: © Gramazio Kohler Research, ETH Zurich and © Renate Weissenbock

All the projects are the result of a performance-based process through which the generative procedure is informed by performative parameters (e.g., structural, environmental). This methodology enables the exploration of innovative formal codes utilizing various typologies of plastics polymers as construction materials. The presence of original digital fabrication tools such as robots implies that the universities involved in the survey are the ones equipped with robotic manufacturing laboratories who work on 1:1 scale projects.

4.2.1 ETH, Gramazio Kohler Research: Robotic Fabrication of Acoustic Brick Walls

The selected case study *Robotic Fabrication of acoustic brick walls*[1] (Fig. 4.2) represents the synthesis of the integration of innovative technologies, from design to production, aimed at creating informed architectures [23]. The study takes especially

[1] In cooperation with: EMPA: Kurt Eggenschwiler, Dr. Kurt Heutschi; Research programme: CTI research project; Collaborators:Max Vomhof (project lead), Dr. Ralph Bärtschi, Thomas Cadalbert, Lauren Vasey, Guilherme da Silva Carvalho, Luis Gisler, David Jenny, Clemens Klein; Consultancy: Jürgen Strauss Industry partner: REHAU Vertriebs AG.

Fig. 4.2 *Robotic fabrication of acoustic brick walls*, 1:1 scale prototype. © Gramazio Kohler Research, ETH Zurich

into consideration the importance of acoustic comfort, particularly of the reverberation phenomenon, in non-dedicated architectural spaces (e.g., residential units, classrooms) and how particular attention on defining the materiality of the surfaces and their geometrical configuration is required. The research focused on this specific topic by designing a performative surface able to solve or improve the reverberation phenomenon thanks to its geometric configuration and material qualities [24].

The starting surface is generated by a grid of points in space and is modified through the management of the coordinates (x, z) of the points themselves to achieve the desired acoustic performance and improve indoor comfort conditions. In addition to solving the acoustic problem, the designed plane has to guarantee structural rigidity and the effective production of the components taking into account all the parameters that describe the fabrication process. The surface designed following the criteria described above is then discretized to determine the individual elements, small plastic bricks. The components consist of a main body shaped hexagonally which defines the length, and a secondary body that assures differentiation in the depth of the surface (Fig. 4.3). In order to obtain high performance through a simple and low-cost system, it was decided to experiment with the use of plastic materials, such as the common ABS polymer utilized in 3D printing. Such material, apart from being sustainable, can guarantee geometric flexibility and ease of assembly. Every single component was realized by a common 3D printing process using a common table-desk machine. Thanks to a design-to-production procedure, the codes that inform the robotic manufacturing process are extracted directly from the geometrical data that describes the performative surface. The robotic arm KuKa KR 150 utilized for fabrication, mounted on a linear axis of 8 m, assembles the various elements in a simple pick and place operation to which an ultrasonic welding device is added to connect the different components and guarantee geometric and structural continuity. For the production process described above, it was necessary to design and prototype a specific tool able to carry the operations in continuity. The realization of a customized end effector allowed to inform the production procedure in such a way as to correct any inaccuracies and tolerances through continuous feedback between the digital model and the manufacturing process. The two prototypes realized were

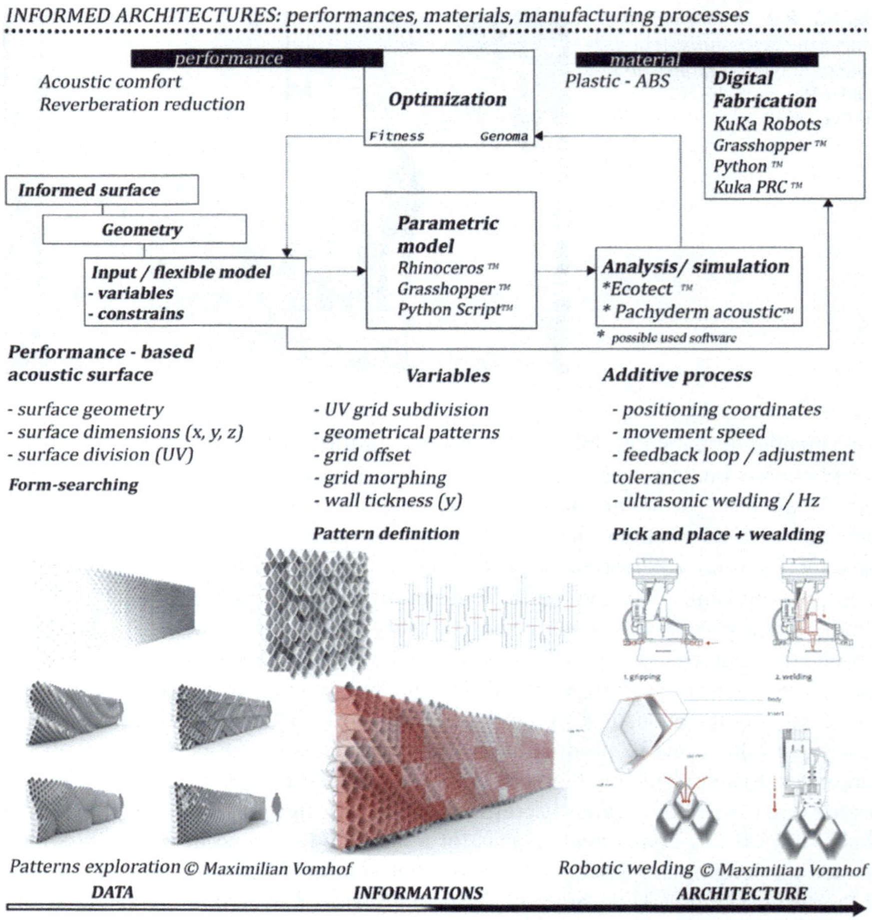

Fig. 4.3 *Acoustic Bricks Wall*: performances, materials, manufacturing processes. A design methodology. © Angelo Figliola. Copyrighted material: © Maximilian Vomhof

tested and verified in order to validate the proposed solutions and to proceed with the comparison with traditional systems. The case study demonstrates how the use of innovative technologies, together with a vision of cyber-physical making, is fundamental in the creation of informed architectures. Performative surfaces generated through a process of geometric optimization, using data-driven strategies, inform the design and construction procedure and therefore the final product. It is evident how, thanks to a meta-project defined by the designer, establishing a series of variable and invariable parameters, it is possible to access a space of design solutions sufficiently wide and continuously fed by the maximization of acoustic performances. The variable parameters are represented by the subdivision of the surface, UV, and the consequent deformation of the corresponding grid points. The material, in the

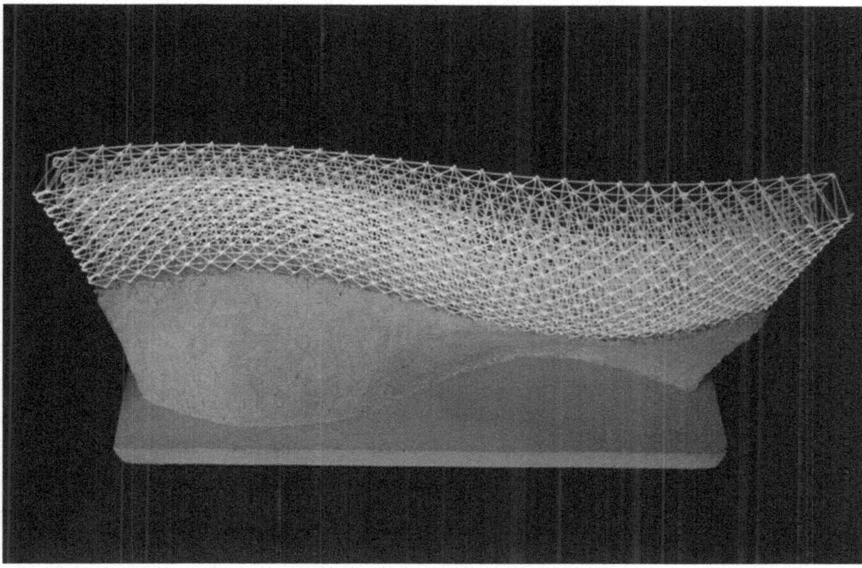

Fig. 4.4 *Mesh Mold*, 1:1 scale prototype. © Gramazio Kohler Research, ETH Zurich

specific case of plastic polymers, informs the design and production process, representing a further design variable. The realization of the responsive surface, composed of a series of customized elements, is achievable due to the construction of an end effector that guarantees the correct positioning of the components as well as the securing of the elements themselves. The EMPA laboratories, technical partner of the project, certified the acoustic performances of the technological system, demonstrating the actual improvement of the executions. From the project analysis, it is possible to observe that the management of the digital process of acoustic performance optimization as well as the fabrication procedure is complex representing a critical aspect of the proposed methodology (Fig. 4.3).

4.2.2 ETH, Gramazio Kohler Research: Mesh Mold

The *Mesh Mold*[2] (Fig. 4.4) case study presents a technological network for concrete structural elements composed of a spatial mesh made of PLA and glass fibers reinforced concrete [9]. The project starts with the analysis of the reinforced concrete structure's construction procedure. This process is characterized by the use of

[2]In cooperation with: Physical Chemistry of Building Materials group (Prof. Dr. Robert J. Flatt), Institute for Building Materials, ETH Zurich; Collaborators: Norman Hack (project lead), Willi Viktor Lauer; Industry partner: Sika Technology AG, Switzerland.

three technological elements: cast containment frames, formworks, metal bars for structural fortification, and concrete.

Starting from this evaluation, the research develops a spatial complex as an open system adaptable to free-form morphologies capable of guaranteeing structural stiffness and stability to the concrete, acting as reinforcement and the formwork at the same time [10, 11]. The creation of the mesh allows to avoid formwork and to utilize the concrete through a spraying process. The optimization procedure can involve two design scales: at the macroscale there is a development of a complex and structurally efficient plane, while at the microscale maximization of the spatial construction (grid structure) according to the variable parameters of the production procedure. There are several inputs to be considered while defining the system. Firstly the pressure of the concrete at the time of laying, based on which the density of the mesh is determined. Secondly, the topology of the mesh which influences the fabrication times, the quantity of material used, and the loads deriving from the iteration between concrete and spatial structure. Compared to the criteria listed above, a series of structural tests were carried out on prototypes measuring 80 cm × 30 cm × 15 cm starting from a double-curved surface and identifying three load points. The tests are repeated by varying the mesh topology to verify the structural behavior, especially by analyzing the parameters related to the deformation, the distribution of loads, the quantity of material utilized, and the duration of the production process. Based on these parameters, the morphology was tested on a 1:1 scale. One of the most complex aspects of the research involved the manufacturing procedure of the spatial structure in PLA. In order to create the mesh, a robotic arm equipped with a special extruder was employed to guarantee freedom and precision of movement in space, following the geometrical coordinates defined and extracted from the digital model. Thanks to the high number of degrees of freedom of the robotic arm, the combination of robot and 3D printing overcome the logic of layer-by-layer extrusion and the predominantly vertical development of the structures. The extruder designed and prototyped during the research is a patent (No. WO/ 2015/034438) and is part of the National Centre of Competence in Research (NCCR) Digital Fabrication research project. The proposed technological system explores the application of an informed strategy in terms of structural data for the realization of a spatial mesh that performs a double task. Specifically, the mesh represents the reinforcement structure for the concrete (replacing the iron bars) and, at the same time, constitutes the formwork to carry out the casting. The application on the 1:1 scale prototype allowed to test the actual applicability of the methodology and process (Fig. 4.5). Despite this, some critical issues persist both in the management of the production and construction procedures, especially in terms of spatial mesh continuity around structural nodes (e.g., wall-slab) considering the dimensions reached by the robotic arm. In the realized prototype the dimensions of the component facilitated the extrusion process, particularly in terms of the z-axis bound to the robot's movement potentialities. The possible implementations concern the development of constructive and assembly strategies able to guarantee the creation of continuous joints of greater complexity and the application of a mobile system that allows the use of the robot in situ. For

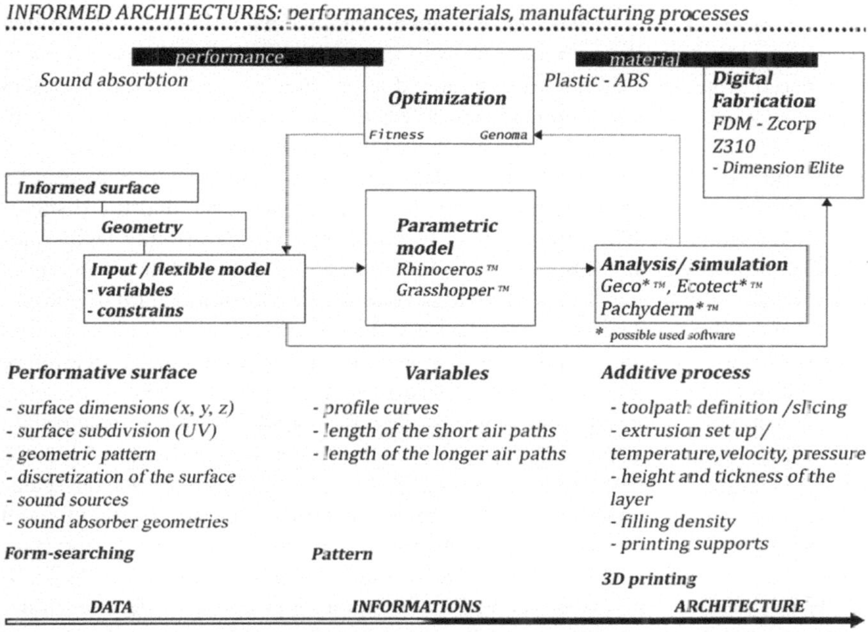

Fig. 4.5 *Mesh Mold*: performances, materials, manufacturing processes. A design methodology. © Angelo Figliola. Copyrighted material: © Norman Hack

this purpose, the researchers are currently developing a material system composed of metal strands for a 1:1 scale employment.

4.2.3 TU Delft: Acoustics by Additive Manufacturing

The proposed case study is part of a research line focused on the application of additive manufacturing in architecture, as a production method useful for the realization of informed technological systems resulting from a process of optimization of acoustic performances [21]. The research studied different types of technological networks able to absorb or diffuse sound by exploiting the potential offered by the computational procedure and by digital fabrication in shape customization. The research specifically investigated these aspects: fractal geometrical systems and complex segmented systems for sound absorption; optimized curved surfaces and material networks with a differential superficial gradient aimed at improved sound wave diffusion. The system chosen for in-depth analysis is characterized by a defined geometry as a passive destructive interface according to four main factors: influence of the geometric configuration on the acoustic performance of the system, on material and production method used, and on economic aspects. In addition to a series of scale prototypes utilized to make measurements and understand the real absorption of sound in relation to the geometric traits of the tube, such as length and diameter, the research proposes a design concept aimed at producing a technological system serving as sound absorber applicable to internal spaces free of interference with external climatic factors [20, 22]. In the methodology proposed for the realization of the technological networks, the digital model is informed by parameters that describe the absorbing system as the definition of the space of influence and the distance between the air paths. The prototypes were made through 3D printing techniques and ABS, a plastic commonly used for additive processes, to exploit the potential offered by a common technology as well as guarantee the customization of components to meet the performative requirements. The analyzed technological system explores the employment of an informed strategy for acoustic data in the creation of a sound-absorbing system for indoor applications. Through the proposed methodology, a string of responsive models was studied for sound diffusion or absorption, maximizing a series of project input parameters. Furthermore, in order to inform the digital procedure, a sequence of tests were carried out on prototypes at different scales with the intent of understanding the phenomenon in terms of geometry and material characteristics. The data collected was utilized for the development of a preliminary concept for the acoustic technological system, which was realized at the Delft Faculty of Architecture. The concept, as well as the physical tests, defines guidelines to support designers in the development of performance acoustic technological networks (Fig. 4.6). Despite the theoretical value of the suggested methodology, some operational problems concerning the management of the parametric process to establish the geometry remain, as well as limitations in terms of reduction of the component's dimensions according to the manufacturing method used. Among the possible implementations (for example robotic manufacture) there is a chance to obtain larger elements and introduce a feedback loop system between the extrusion process, material, and environmental parameters. Research on new production technique goes hand in hand with studies on new materials to increase performance through optimized geometric criteria and with the digital procedure implemented through genetic maximization algorithms.

4.2.4 TU Delft: Additive Manufacturing for Daylight. Toward a Customized Shading Device

The proposed case study (Fig. 4.7) is part of the research path on the application of additive manufacturing technologies to produce optimized technological systems informed by environmental performances aimed at improving indoor and outdoor comfort [14]. The study focuses on the design of a customized shielding system based on a series of parameters that describe the geometric and material characteristics of the system, as well as the production process. The final decision consists of mediation between the choices of the designer and those of the end user who can make the

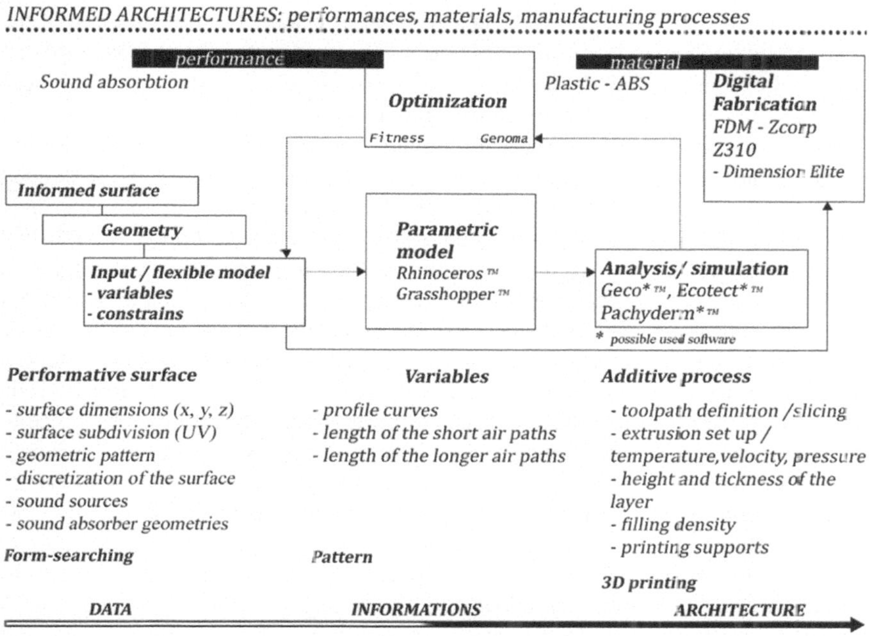

Fig. 4.6 *Acoustic by Additive Manufacturing*: performances, materials, manufacturing processes. A design methodology. © Angelo Figliola

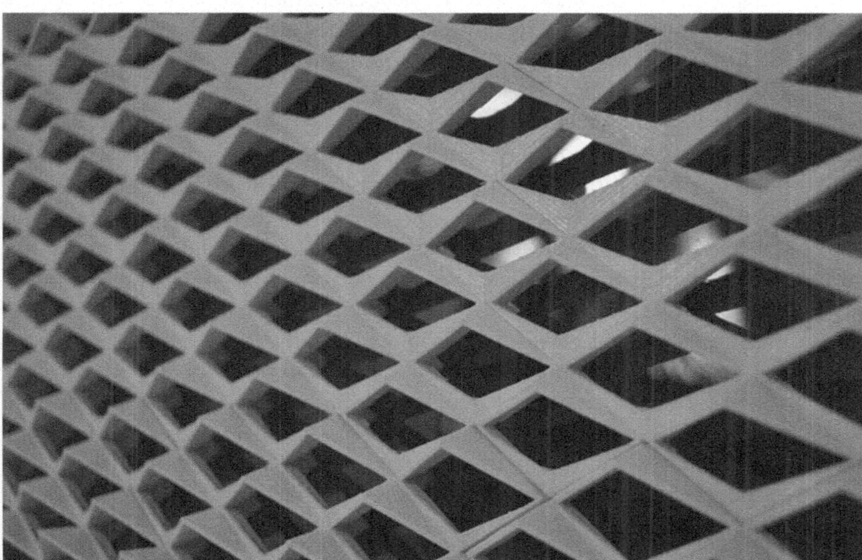

Fig. 4.7 Patterns exploration © Lemonia Karagianni

appropriate changes through a dedicated web employment. The construction of the 1:1 scale shading apparatus prototype is the result of a consequential procedure that involved different levels of detail, from macro- to mesoscale, in which user's choices play a fundamental role in the meta-project defined by the designer.

The global parameters relating to the climate zone, site orientation, dimensions of the space where the device is to be applied, and the geometric pattern, were defined at the macroscale, while elements and assembly systems of the technological units were studied at the local scale. Within the space of design possibilities, established by the parameters mentioned above, environmental performance such as daylighting, shading, and glare analysis, informed the entire process, guaranteeing an optimized morphology in terms of user-defined requirements. In order to realize the prototype, the material system was studied in such a way as to warrant the performances necessary for its real-scale application. The criteria considered in the choice of the material through the CES Edupack 2014 platform were: recyclability, resistance to UV and water, temperature range, and resistance to fire. At the end of the procedure, polyethylene terephthalate (PET) was selected. PET is available for FDM 3D printing technology, and it is inexpensive. Once the geometry and the components were determined, it was necessary to define a system of customized joints able to assure structural stability for horizontal loads. The idea of creating an online platform (*DesignyourShatter.com*) which allows the user to delineate the morphology based on his/her performance needs, introduces a new concept of mass customization and production technologies' democratization [15]. The proposed technological network explores the employment of an informed strategy, founded on environmental data, for the construction of a shading system for architectural façades. The proposed methodology (Fig. 4.8) defines a meta-project in which a string of geometric parameters informed by environmental analyses determine the final morphology achievable through a 3D printing process. The design procedure involves the study of variable geometric patterns, material used, and assembly system. The technological system is suitable for applications in multi-layer enclosures, both in new constructions and in existing buildings' retrofitting operations. The utilization of PET ensures adequate resistance to UV rays and water but has no resistance to fire. By observing the proposed case study, it is possible to hypothesize a series of implementations. Firstly, in the use of innovative materials able to assure performance properties for compliance with ISO standards on shading devices. Secondly, in terms of the development of research on translucent materials for 3D printing. Thirdly, in the definition of joints with greater structural executions capable of withstanding vertical loads applied. Lastly, in terms of the generation of a digital parametric process aimed at guaranteeing greater control in terms of environmental and thermal criteria. Furthermore, the case study introduces a new concept: the creation of a web platform where the user can intuitively inform the procedure through specific data about the application context and print the device with a 3D printer directly at home.

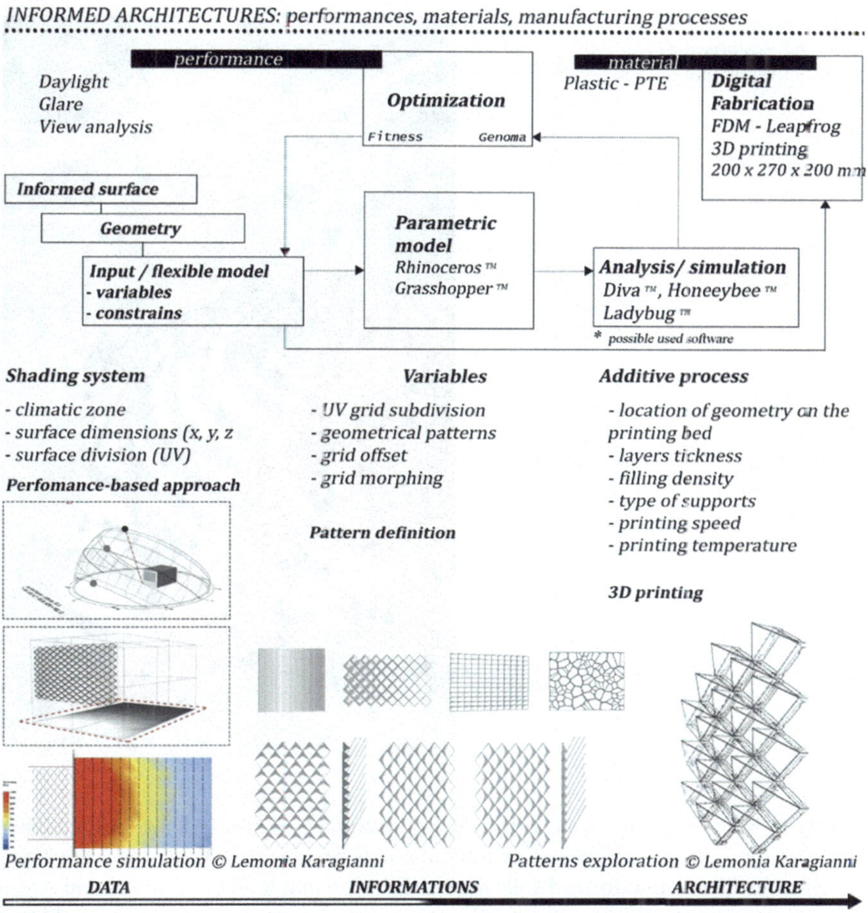

Fig. 4.8 *Additive Manufacturing for daylight*: performances, materials, manufacturing processes. A design methodology. © Angelo Figliola

4.2.5 *MIT, Massachusetts Institute of Technology: Robotics-Enabled Stress Line Additive Manufacturing*

The case study examined suggests an operative methodology for the creation of 2.5D informed surfaces in which the structural performances are increased and optimized through the deposition of material along main stress lines identified following simulations and analysis (Fig. 4.9). Consequential phases of development characterize the process. A first phase, in which the criteria is defined and describes the design domain—the minimum and maximum volume of the geometry, arrangement, and

Fig. 4.9 *Robotics-Enabled Stress Line*, prototype. © Kam-Ming Mark Tam

quantity of support points—as well as the consequent creation of the surface catenary. A second phase, where the obtained plane is reviewed. A third and last phase, where the surface is maximized following the design and load conditions, and stress lines are generated utilizing distribution rules to correct and eliminate problems deriving from the density and placement of the latter on the plane.

The strain energy optimization method is used to manage the stress distribution. Such a technique determines a discretization process based on symmetric elements which compose the shell surface. The proposed methodology, (*Stress Line Additive Manufacturing*) has the objective of interconnecting a method of structural analysis with digital manufacturing to define new design possibilities for informed structures [17]. The prototype is the combination of different production processes: milling to build a support at the base and avoid problems with anchor points, and robotic 3D printing to distribute material along the structural stress lines. The employment of a robotic arm and a customized end effector allowed to obtain flexibility in space and to assemble the material based on the desired performances. The prototypes were subjected to structural checks for normal stress and horizontal displacement by differentiating the type and load conditions. The project *Robotic-Enabled Stress Line Additive Manufacturing* investigates an innovative fabrication methodology, extrusion of thermoplastic polymers for subsequent layers, and *FDM* for the manufacturing of optimized surfaces where structural performances become a design input that determines the final morphology. The design of the prototypes follows a maximization process that acts first on the surface based on the variable parameters that characterize the meta-project and, subsequently, on the distribution of stress lines, whereby the orientation determines the direction of the extrusion path. The project methodology (Fig. 4.10) and the prototype demonstrate how it is possible to increase structural performances through innovative digital technologies, both in the design and construction phase. Furthermore, this approach to the problem of structural optimization presents numerous critical issues derived from a possible employment to the architectural scale for structural purposes. The limits derive from the geometrical dimensions of the buildable technological units and the material utilized. Even

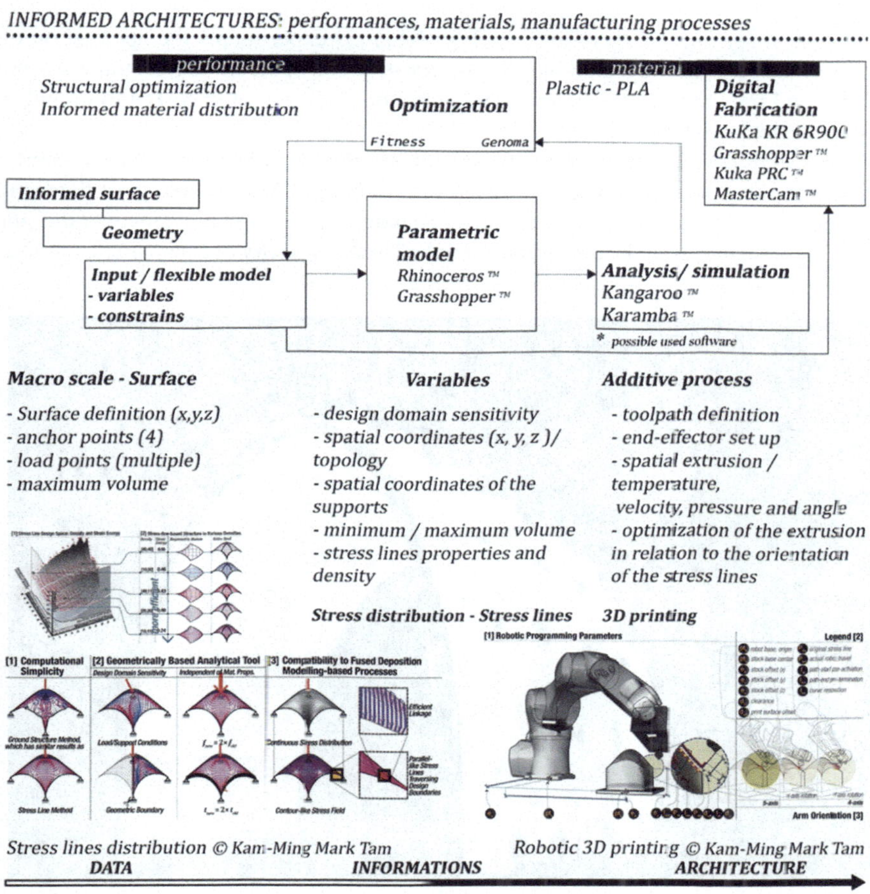

Fig. 4.10 *Robotics-Enabled Stress Line*: performances, materials, manufacturing processes. A design methodology. © Angelo Figliola

if the robotic arm can guarantee greater freedom of movement and can rely on additional external axes (e.g., track, rotating table), it is still bound by the environment in which it moves. While, for what concerns the material, the polyatomic acid (PLA) is not suitable for structural applications as it does not assure the right performance parameters founded on codes and regulations (e.g., ISO). Considering the limits of the procedure described above, the possible implementations of the methodology should involve the development of new material systems and the improvement of the production process, as well as inquiries on the structural tests' results which presented anomalies compared to the trials on standard shell surfaces.

4.2.6 Kent State University: Solar Bytes Pavilion

The case study *Solar Bytes Pavilion* (Fig. 4.11) [3] experiments the application of small-scale 3D printing techniques for the construction of technological units assemblable and transformable into an architectural body through the design of complex technological systems [25].

The pavilion is the result of an investigation in the field of subdivision and consequent paneling of an initial surface, useful for defining the morphology of the technological unit and optimizing its geometry founded on the daily movement of the sun [16]. The executions of the prototype are linked to the capacity of every single module

Fig. 4.11 *Solar Bytes Pavilion*, design concept. © Brian Peters

to capture the largest amount of natural light and solar radiation based on their position relative to the sun. The incident solar energy is stored during the day to supply a LED light inserted in the technological unit, which guarantees artificial lighting at night [12]. Through this digital process, each module is customized based on the data that informs the project. The tailoring of every single module makes it necessary to design a fast and structurally efficient interlocking joint system which assures structural stability and ease of assembly. Each technological module is built with a medium-size Kuka Robot by a 3D printing procedure through a customized extrusion process that ensures the correct distribution of the material, as well as a quick grip of the latter. In order to set up an efficient production procedure, it was necessary to carry out a series of experiments that allowed to analyze, with a thermal imaging camera, the complexities derived from the additive process, and modify accordingly the parameters that control the extrusion, mainly temperature, pressure, and speed of movement. The examined case study determines an operative strategy (Fig. 4.12) that explores the possibilities offered by small dimension 3D printing for employments to the architectural scale, dealing with environmental and energy sustainability [18]. The creation of energy linked to the performance of the individual technological units represents an added value of the entire process and opens up further implementations in terms of the technologies utilized, the materials, and the possible spatial configurations, aiming at the increase of the performance mentioned above. Even if the tests carried out on the modules demonstrate a good mechanical resistance, the methodology presents a criticality related to the components used. Although excellent for prototyping, the plastic polymer employed, specifically ABS, does not meet some of the execution requirements for their application to the architectural scale, such as fire resistance. The presence on the market of bioplastics, with adequate performances for utilization on the architectural scale, makes the procedure and the product extremely suitable for architectural employments.

4.2.7 Graz University of Technology: D-FORM

The *D-Form* [26] case study presents a series of façade panels (Fig. 4.13) whose variation is determined by performance used as a design input based on material, geometric patterns, and robotic fabrication process. Informed geometries resulting from the procedure described above allow defining an adaptive relationship between geometry and environmental parameters, such as air permeability, natural ventilation, lighting, and solar radiation, as well as specific points of view. The experimentation utilizes acrylic glass panels, plexiglas, manufactured through two consequential processes resulting from the combination of different digital fabrication techniques: firstly, through a numerical control laser cutting machine (CNC), the geometric pattern chosen in relation to the desired performances is engraved, while in the next phase the pattern is modified through a robotic thermoforming procedure. The plexiglas panel is mounted in wooden frames with variable geometry. Thanks to the presence of a deforming element and the possibility to move the panel in space

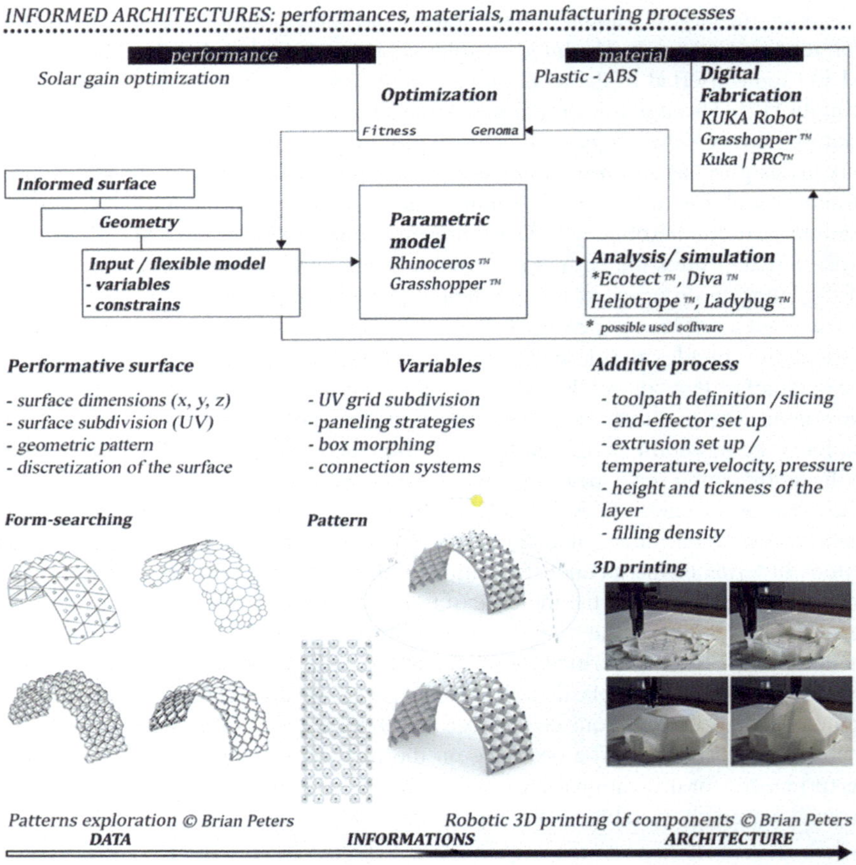

Fig. 4.12 *Solar Bytes Pavilion*: performances, materials, manufacturing processes. A design methodology. © Angelo Figliola

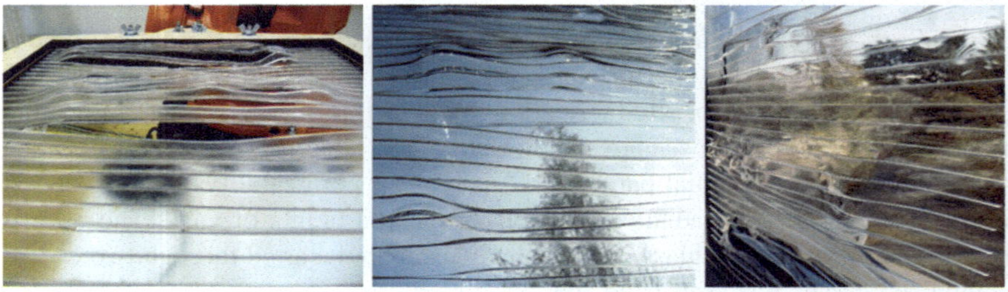

Fig. 4.13 *D-FORM*, façade panel prototype. © Renate Weissenböck

with precision, the motif is modified to obtain diverse geometric configurations, following the desired performative parameters. The robotic arm, wherein the panel is attached, can move quickly and precisely in the processing space while the whole procedure is simulated in a digital environment, calculating the heating and cooling times necessary to guarantee the thermoforming process and the correct deformation of the surface. Specifically, the research investigates four geometric patterns of tessellated surfaces: rhomboidal, triangular, continuous, and segmented geometries which constitute a catalog of solutions to which the designer can refer by defining the meta-project.

The proposed prototypes have as objective the definition of an operative methodology for designing and fabricating façade panels employable in multi-layer envelope systems. The *D-Form* project and the methodology adopted demonstrate how the combined use of computational design and diverse digital manufacturing methods allows to design and produce informed geometries characterized by a perfect correspondence between the digital model and the physical prototype [26]. The 1:1 scale prototypes show all the performance parameters required for architectural application in multi-layer facades, despite the numerous criticalities deriving from the combination of different fabrication processes. In this regard, although the robotic manufacturing procedures (Fig. 4.14) can guarantee high precision, the management of criteria demands a series of complex digital simulations to ensure the correct success of the process in terms of deformation of the geometric pattern obtained through thermoforming (Fig. 4.15). The choice of the geometric motif can be related to the required performances based on specific climatic conditions and a certain programatic function. Its morphology can be optimized utilizing a genetic algorithm able to determine informed and adaptive geometries and define an improved and extended space of design possibility. The four proposed façade panels can be implemented through the use of sustainable materials, such as bioplastics or smart materials, as well as by studying intelligent systems connecting the panel's primary and secondary structure.

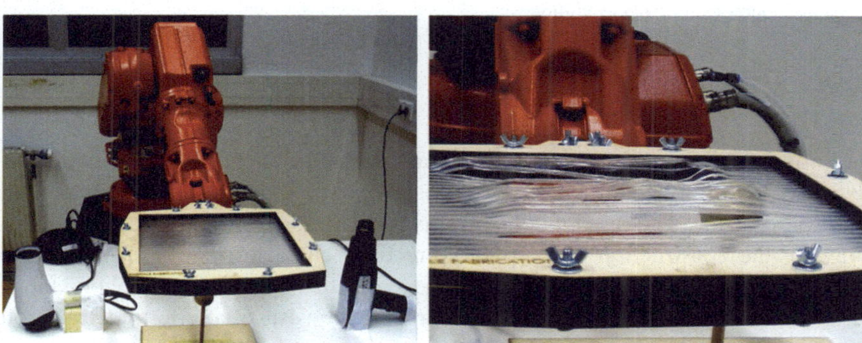

Fig. 4.14 *D-FORM*, robotic fabrication process. © Renate Weissenböck

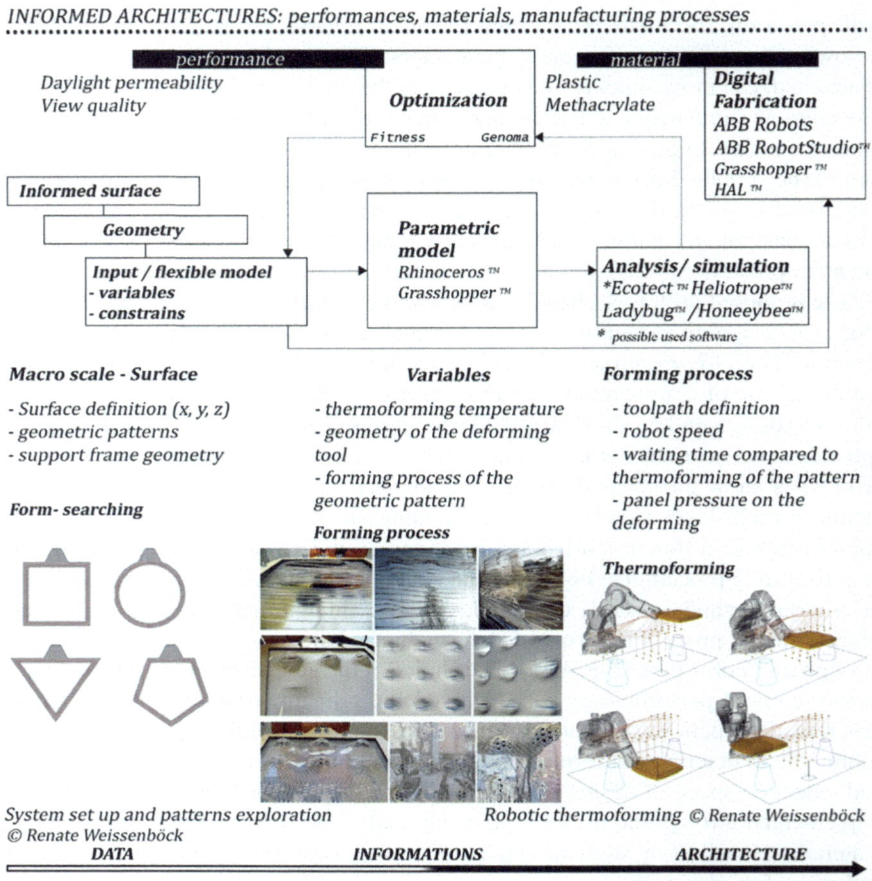

Fig. 4.15 *D-FORM,* Renate Weissenböck: performances, materials, manufacturing processes. A design methodology. © Angelo Figliola

4.2.8 AA School: Osteobotics[3]

The proposed case study [2, 3] presents an experimental prototype in which an advanced and informed structural system is expressed through a robotic manufacturing process [4]. The research objective is the definition of an operative methodology for the construction of temporary erections characterized by an innovative structural concept, the use of sustainable and synthetic materials and a robotic fabrication process with in situ assembly. The structure consists of a series of tetrahedral

[3]DRL Team: Mattia Santi, Soulaf Aburas, Maria Paula Velasquez, Giannis Nikas. Shajay Bhooshan Studio—Architecture and Urbanism MArch (DRL)—Architectural Association School of Architecture.

joints obtained by the spatial development of a starting topology generated by the protrusion of the faces (Fig. 4.16). The tetrahedral joints are connected by a binder material (thermoplastic polymers) through a custom-made extruder and two robotic arms that work collaboratively. The procedures described above have been designed to be implemented on the project site due to a mobile network able to place the robots directly in situ [1].

The surveys concerned both the study of the material system, with the search for the optimal fabrication set up especially referring to extrusion parameters and the robotic manufacturing process concerning the coordination of the various production phases involving collaboration between man and machine. The fabrication procedure involved the coordination of two robotic arms that perform the following actions: through a custom end effector, the tetrahedral joints deposit the plastic material through extrusion based on the geometry defined by the digital model. The next stage consists of the plastic polymer extrusion and therefore the creation of the linear structural connection between the tetrahedral joints. A fixing lacquer is utilized to stabilize and guarantee structural stiffness following the evolution of the fabrication process. The construction of a tetrahedral joint, with four connections, requires a customized tool and 500gr of extruded plastic polymers at a temperature of 60 °C.

Fig. 4.16 *Osteobotics*, structural system prototype. © Mattia Santi

The proposed technological system explores the application of an informed data-driven strategy for structural, material, and manufacturing procedure, aimed at the realization of an optimized construction composed of discretized linear elements connected by tetrahedral joints. The case study experiments innovative structural principles which are the use of an eco-friendly material and an automated additive manufacturing system to be employed directly in situ. The design methodology (Fig. 4.17) can be applied for the construction of temporary structures, as well as for vertical partitions in multi-layer envelopes, and represents a sustainable process thanks to the material used and the assembly operation taking place in situ and with a completely dry proceeding. The plastic substance, which makes up the connection system between the tetrahedral joints, is eco-friendly and recyclable: due to these characteristics each structural element can be utilized several times to build

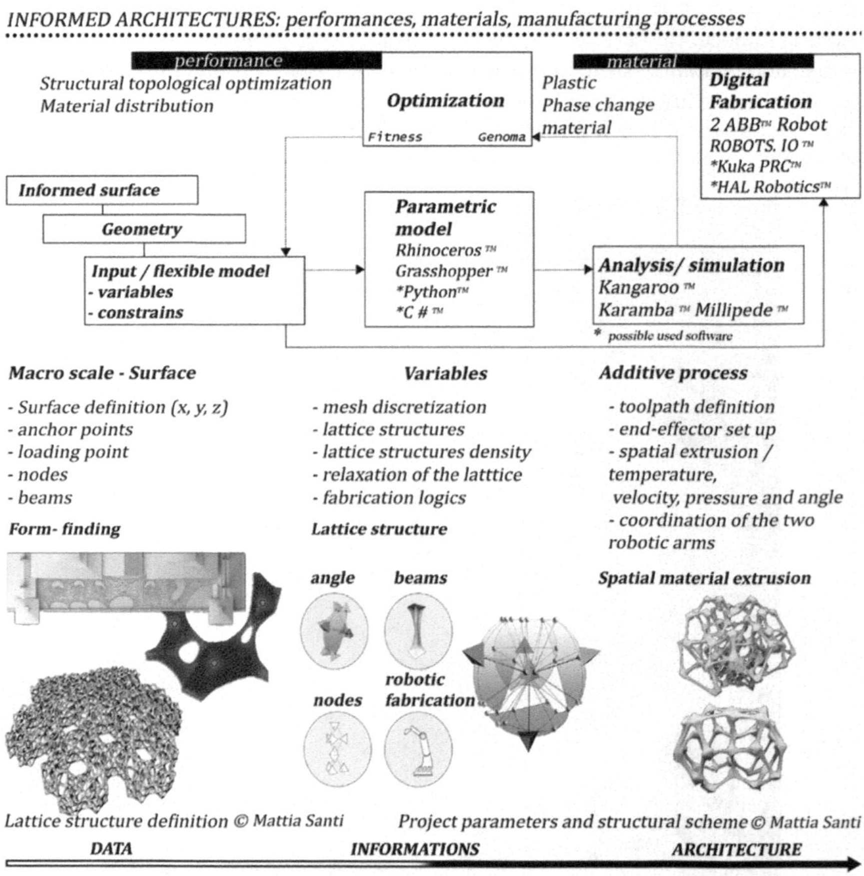

Fig. 4.17 *Osteobotics*: performances, materials, manufacturing processes. A design methodology. © Angelo Figliola

new structures. Although bioplastic represents an eco-friendly material, it does not currently guarantee the requirements necessary for a structural application. This constitutes one of the limits of the execution of this methodology. Another critical aspect is represented by the management of the digital workflow for large-scale architectural applications, which is difficult due to the complexity of both the designing and fabricating processes. Possible implementations could derive from the study of fiber-reinforced materials systems, introducing a new step in the production operation, to assure structural stability and to meet the required regulatory conditions. By applying this methodology, the dimension of the structural apparatus turns out to be bound to the robot used and to the proposed mobile manufacturing system.

4.2.9 Tongij University and Archi-Union: Light-Vault

The case study *Light-Vault* represents the result of a collaboration between a university research center and the architectural firm Archi-Union [27]. The pavilion can be configured as methodological experimentation that starts from the definition of an optimized technological unit and follows its transformation and assembly process aimed at obtaining an architectural organism. From the methodological development of the concept of informed architecture, the investigation leads to the illustration of a digital workflow based on the implementation of genetic algorithms in the design procedure. The latter leads to the definition of an architectural solution that uses structural and environmental performative parameters as generative elements. This operation is informed not only by the performance criteria mentioned above, but also by those corresponding to the material utilized, the production process, and the assembly of various technological units, defining a design strategy informed by data. Thanks to the use of parametric software, the global shape modeled as a catenary surface is maximized and subdivided according to the size of each technological unit. The panels that fill the informed surface present openings which are optimized based on environmental parameters and a double-curved plane. Common materials such as blocks of expanded polyurethane were utilized to build the prototype. The latter can guarantee simplification and speed during robotic fabrication through CNC wire cutting. The material used allows modulating the individual block's porosity by inserting the heated wire for the cutting operation into the volumetric center of the block, maintaining an edge structure to permit the assembly of the various units. Block processing is one of the most laborious tasks compared to the limited amount of time needed for assembly, which takes place in less than 24 h and with the employment of five non-specialized people. The case study analyzed (Fig. 4.18), presents an architectural organism informed by performative data related to the entire design operation, from the morphological generation to the connection of the components. The complexity of the fabrication process and the management of the design parameters make it difficult to scale the product and repurpose it as an architectural organism, while it is particularly suitable for the realization of opaque vertical partitions and multi-layer envelopes, as it serves as an optimized and self-supporting technological

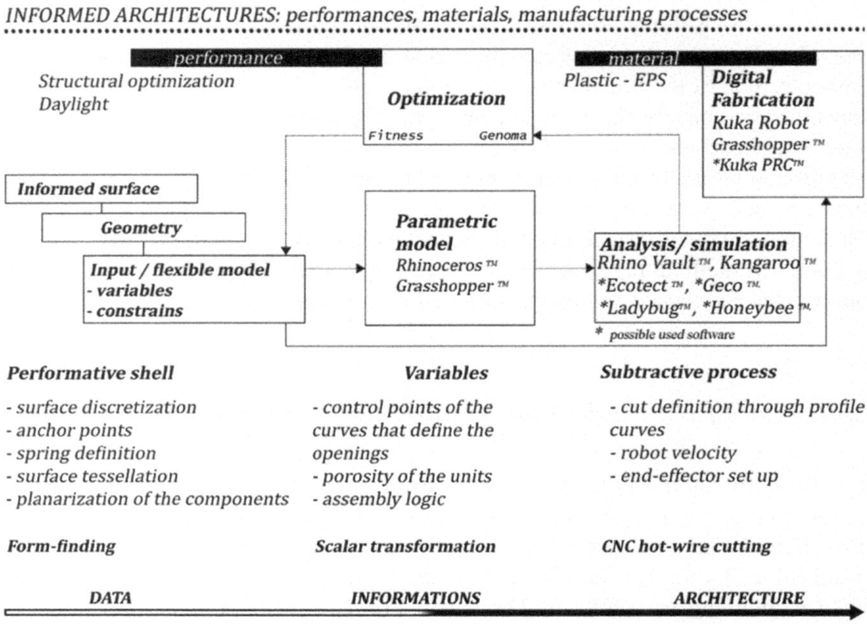

Fig. 4.18 *Light-Vault*: performances, materials, manufacturing processes. A design methodology. © Angelo Figliola

system. In this type of operation, the choice of the material is a crucial aspect of the production process: the need to have a frame that facilitates assembly, imposes the determination of the cut starting point outside the geometry itself, establishing a cut in the external frame that interrupts geometric continuity and undermines overall structural balance. Furthermore, the hot wire cutting production procedure is particularly complex with hard substances, such as stone and marble, which would allow greater structural stability as well as direct application in the construction sector. On the contrary, the use of EPS blocks as a technological unit, although easy to manage, requires further research for the use of finishing materials, such as cementitious resins or plastics to be employed with thermoforming processes, able to guarantee structural rigidity and to ensure a good resistance to atmospheric agents. In addition to the studies on materials, further implementation concerns the tool necessary for cutting operations through the introduction of sensors capable to vary speed and temperature according to the components utilized. Overall, the operation of customizing the technological units and the assembly process is fast, efficient, and able to maximize material use.

4.2.10 *The University of Sydney: TriVoc*

The case study analyzes the acoustic phenomenon of sound diffusion based on the geometry of space and technological devices. The research exploits digital computational and robotic fabrication as useful tools to define an operating methodology able to link the digital model, the simulation of the acoustic phenomenon, the optimization of performance, and the production process (Fig. 4.19). The focus of the investigation is the study of the diffusion of sound in ellipsoidal geometric spaces in which the presence of two focal points and different heights generates a diffusion of a peculiar sound for monofocal circular geometries.

The relationship between acoustic phenomenon and geometry is governed by a mathematical equation capable of generating distinct spatial configurations by diversifying the criteria that describe the algorithm in order to achieve the desired performance and to maximize the morphology. Due to the possibilities offered by the computational process, the spatial configurations described by geometric coordinates obtained through auditory simulation are transferred to the 3D modeling software. The procedure allows interconnecting the 3D optimized model to the fabrication operation, considering the variable parameters and constraints that characterize it. The acoustic simulation generates different spatial solutions described by a series of Cartesian coordinates that constitute the meta-project, while the direct link between geometry and manufacturing process introduces a further variable in the definition of the space of design possibilities. A subtractive procedure applied to a solid block of plastic material produces the prototype of the informed surface. Criteria such as the type of robot used, the end effector, and the maximum and minimum dimensions that describe the working area of the robotic arm inform the final geometry. After a sequence of tests, EPS was chosen as the prototype material due to its ease of processing and efficiency in terms of manufacturing method and auditory properties. The *TriVoc* project takes into examination the relation between simulation, morphological optimization, and production process, considering the latter as a variable able to define the space of possibilities within which the designer can make his choice. While in similar cases the focus is on simulation and spatial maximization, this

Fig. 4.19 *TriVoc*, robotic milling process. © Dagmar Reinhardt

research project considers the construction procedure as a design input to ensure the opportunity of realizing the geometry based on the chosen manufacturing method [19]. The proposed methodology (Fig. 4.20) can be applied to the creation of responsive acoustic internal partitions, and it can be optimized according to various required performances, as well as for tectonic solutions in which structure and partition are not distinct elements. The critical aspects of the operation described above derive from the management of the computational workflow and the consequent transposition of the digital prototype into physical space through robotic fabrication. The interoperability between different modeling, analysis, and manufacturing software is a potential for the design process, but at the same time, it makes the management of the

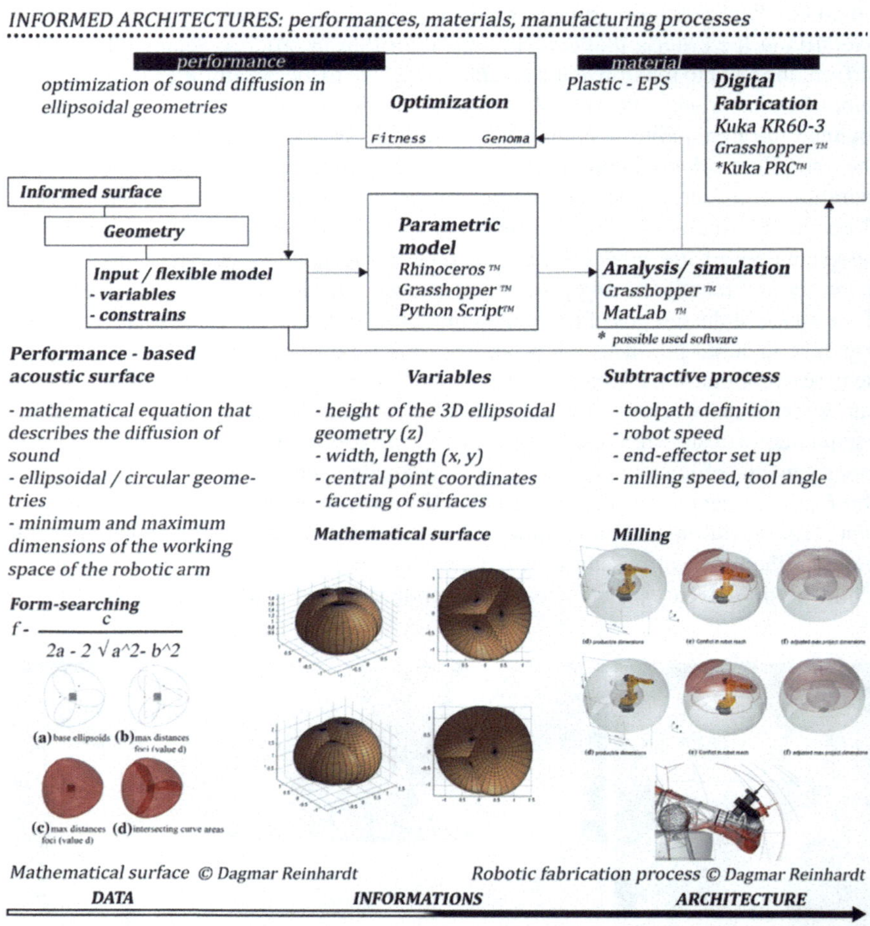

Fig. 4.20 *TriVoc*: performances, materials, manufacturing processes. A design methodology. © Angelo Figliola

Fig. 4.21 Informed porosity, 1:1 scale prototype. © Sina Mostafavi

entire supply chain complex, especially when dealing with non-standard geometries. An important feature needing improvement in the definition of an operation directed at creating a 1:1 scale sample is the materiality of the systems to ensure structural stability and acoustic performance at the architectural scale. A further implementation can be derived from the discretization of surfaces realized through the assembly of parts rather than by subtraction of matter from solid, single material blocks.

4.2.11 TU Delft: Informed Porosity

The *Informed Porosity* case study (Fig. 4.21) is part of a research line that investigates the phenomenon of porosity for the creation of informed structures, in this case a pavilion, as a result of an optimization process and realized through robotic manufacturing [13]. The construction of the 1:1 scale prototype is the outcome of a consequential operation that involved different scales of analysis such as macro-, meso-, and microscale, and applies a strategy informed by data related to structural, energetic-environmental, acoustic, and topological performance parameters.

At the macroscale, the research starts from the definition of a meta-project which delineates the morphology of the architectural organism before developing and employing the procedure to the meso- and microscale and determining the portion of the pavilion analyzed. The computational process of multi-objective optimization linked the data with the geometrical traits of the surface: the openings on the surface are established by the simulation of dynamic environmental phenomena such as daylighting, while the distribution of the loads along the plane determines the variation of the openings through the material thickness. Finally, the curvature of the surface and the acoustic performances define the tessellation and the density alteration of its units. Informed geometry obtained by a maximization procedure was created through a subtractive milling operation, applied on a block of EPS. With the employment of this methodology, every component of the architectural organism is highly customized as it responds to specific inputs in terms of its spatial location and this requires a production process able to guarantee the realization, limiting time and costs. The case study explores the principle of porosity applied to different scales of investigation and a multi-objective optimization process in which performance data informs the final morphology. The 1:1 scale prototype realized, represents the result of the application of the methodology described above as an explanation of the relationship between execution and geometry: an informed surface able to maximize the use of the material and assure adequate structural, environmental, and acoustic performance founded on the functional program established in the meta-project. The proposed methodology (Fig. 4.22) can be employed for the creation of non-standard

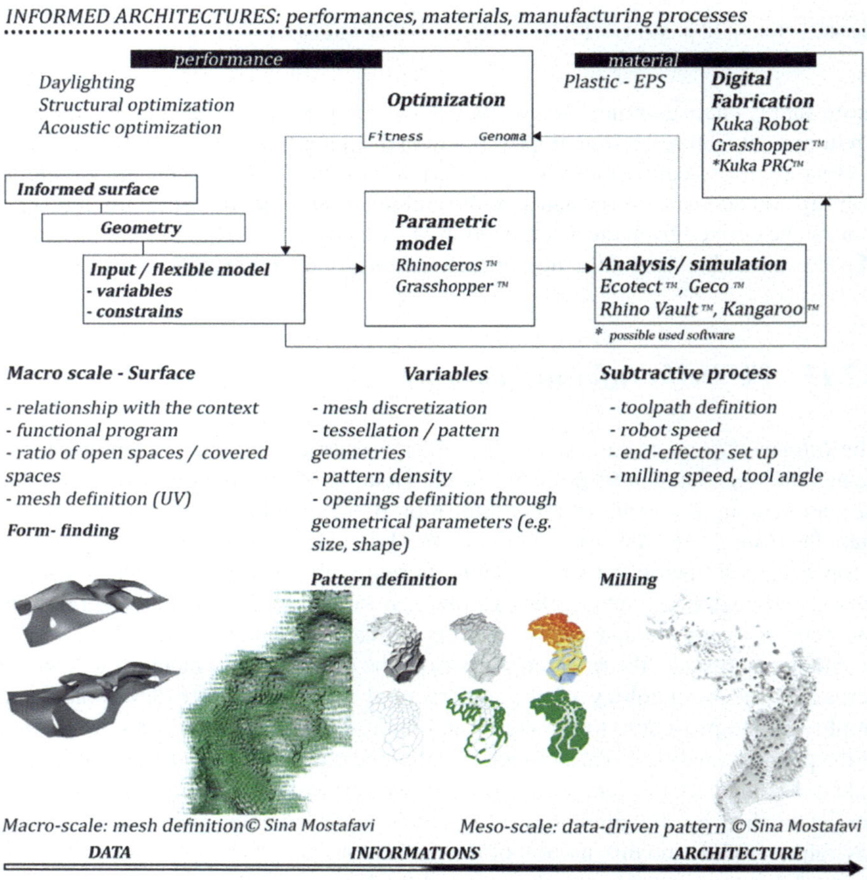

Fig. 4.22 *Informed Porosity*: performances, materials, manufacturing processes. A design methodology. © Angelo Figliola

structural components and for the construction of optimized vertical partitions. multi-layer envelope, where responsiveness is required based on variable and differentiated inputs. The critical aspect of the methodology consists in the dimensions of the robot and the geometric coordinates defining the working space and the manufacturing method adopted (milling) due to its poor efficiency in terms of production times. A possible further implementation can be derived from the development of a constructive strategy that provides for the discretization of the components, considering the working space of the robot as a dimensional limit for the individual units that will be assembled in the following phase.

4.3 Conclusion: Informed Plastic Systems

The use of thermoplastic polymers plays an important role in the framework of research on the topic of informed architecture, especially in the prototyping phase of technological components. Compared to the common utilization of plastic materials in the AEC sector, mainly for the production of technological elements, the combination of computation and digital fabrication and the development of thermoplastic polymers favored the generation of a hybrid testing ground at different project scales, from micro- to macroscale. The combination of plastic polymers, additive processes, and robotic manufacturing has opened up an innovative scenario for the design of informed technological systems created by maximizing the performance parameters related to structural and energy-environmental aspects. Furthermore, the procedure described above allows optimizing the employed material resources according to the informed design processes through diverse digital fabrication methods ranging from 3D printing to subtractive manufacturing and smart assembly. The relative simplicity of processing, the cost-effectiveness of the raw material, and its ductility facilitate the prototyping of design solutions in the preliminary stage of research, to then move on to the production of selected services. To date, the use of such materials is relegated to the prototyping phase for the external vertical partitioning technology networks due to the physical and mechanical properties of the substance in terms of parameters on fire resistance and atmospheric agents prescribed by EU codes and normative. For indoor applications, the case studies analyzed in the chapter showed how the design methodology is ready for employment in the AEC sector, especially in terms of technological systems realized through 3D printing processes. Another innovation introduced by the case studies concerns the possibility of customizing the technological components in line with specific performance criteria identified by the user and later developing them directly on-site due to diffusion and democratization of printing devices. Future research generations involve simplifying computational performance optimization processes and creating digital design-to-production protocols to simplify component design and user interaction.

References

1. Aburas S (2016) Automated robotic fabrication for temporary architecture. http://www.designboom.com/architecture/robotic-fabrication-osteobotics-aadrl-05-10-2016/. Accessed 24 May 2016
2. CasaClima (2016) Osteobotics nuova vita per le architetture temporanee. http://www.casaeclima.com/ar_27118__Osteobotics-nuova-vita-per-architetture-temporanee.html. Accessed 28 July 2016
3. Design.Lab.Workshop (2015) Design lab workshop's solar bytes pavilion reacts to solar exposure. http://www.designboom.com/architecture/design-lab-workshop-solar-bytes-pavilion-cleveland-ohio-01-15-201\5/. Accessed 20 June 2016
4. DRL (2016) Osteobotics. http://drl.aaschool.ac.uk/portfolio/osteobotics-3/ Accessed 17 May 2016
5. Engelsmann S, Spalding V, Peters F (2010) Plastics: In Architecture and Construction. Birkhauser, Basel
6. Gramazio F, Kohler M, Willmann J (2014) The robotic touch: how Robots change architecture. Park Books, London
7. Gramazio F, Kohler M (2014) Acoustic Brick. http://gramaziokohler.arch.ethz.ch/web/e/forschung/229.html. Accessed 22 July 2016
8. Gramazio F, Kohler M (2014) Mesh Mould. http://gramaziokohler.arch.ethz.ch/web/e/forschung/221.html. Accessed 29 July 2016
9. Hack N, Lauer WW (2014) Mesh-Mould. Made by Robots: Challenging Architecture at a Larger Scale. Wiley, London, pp 44–53
10. Hack N, Willi W.L, Gramazio F, Kohler M (2014) Mesh Mould: differentiation for enhanced performance. In: Gu N, Watanabe S, Erhan H, Hank Haeusler M, Huang W, Sosa R (eds) Rethinking Comprehensive Design: Speculative Counterculture, Proceedings of the 19th International Conference on Computer Aided Architectural Design Research in Asia, CAADRIA 2014, Hong Kong, pp 139–148
11. Hack N, Willi WL, Gramazio F, Kohler M, Langenberg S (2014) Mesh Mould: robotically fabricated spatial meshes as concrete formwork and reinforcement. In: Gramazio F, Kohler M, Langenberg S (eds) Fabricate: Negotiating Design & Making. Gta-Verl, Zurich, pp 224–231
12. Halterman J (2015) The 3D Printed Solar Bytes Pavilion Records the Day and Plays It Back at Night. https://3dprint.com/36730/solar-bytes-pavilion/. Accessed 20 June 2016
13. Hyperbody (2016) Ex25 D2RP—Performative Hybridity. http://ex25.hyperbody.nl/index.php/Msc3workshop:ex25d2rp. Accessed 29 July 2016
14. Karagianni L (2015) Additive manufacturing by daylight: towards a customized shading device [designyourshutters.com]. Master thesis, TU Delft Faculty of Architecture
15. Karagianni L, Turrin M, Knaak U, Hordijk T (2016) Additive manufacturing by daylight: towards a customized shading device [designyourshutters.com]. In: Ramtin A, Chronis A, Hanna S, Turrin M (eds) Proceedings Simaud 2016. UK, London, pp 69–76
16. Maggioni L (2015) Il solar byte pavilion segue il sole verso il futuro dell'architettura. https://replicatore.wordpress.com/2015/04/04/il-solar-byte-pavilion-segue-il-sole-verso-il-futuro-dellarchitettura/. Accessed 20 June 2016
17. Ming K, Tam M, Coleman JR, Fine N, Mueller CT (2016) Robotics-Enabled stress line additive manufacturing. In: Reinhardt D, Saunders P, Burry J (eds) Robotic Fabrication in Architecture, Art and Design 2016. Springer International Publishing, pp 351–361
18. Peters B (2016) Solar bytes pavilion—3D printed interlocking modules. In: Reinhardt D, Saunders R, Burry J (eds) Robotic Fabrication in Architecture, Art and Design 2016. Springer International Publishing, pp 327–337
19. Reinhardt D, Cabrera D, Niemelä M, Ulacco G, Jung A (2014) TriVoc—Robotic manufacturing for affecting sound through complex curved geometries. In: Mcgee W, Ponce de Leon M (eds) Robotic Fabrication in Architecture, Art and Design 2014. Springer, New York, pp 163–179
20. Setaki F (2016) Acustic by additive manufacturing. http://www.slideshare.net/Merford/foteini-setaki-acoustics-by-additive-manufacturing Accessed 12 May 2016

21. Setaki F, Turrin M, Tenpierik M, Van Timmeren A (2014) Acoustic absorbers by additive manufacturing. J Buil Environ 72:188–200
22. Setaki F (2016) Acoustic by additive manufacturing. Master thesis, TU Delft Faculty of Architecture
23. Vasey L (2016) Acoustic brick. http://www.laurenvasey.com/#!gramaziokohleracousticbrick/cjhs. Accessed 22 July 2016
24. Vamhof M, Vasey L, Bräuer, S, Eggenschwiler, K, Strauss, J, Gramazio, F, Kohler, M (2014) Robotic Fabrication of Acoustic Brick Walls. In: Gerber D (ed) Proceedings of the 34th Annual Conference of the Association for Computer Aided Design in Architecture (ACADIA). Roundhouse Publishing Group, Los Angeles, pp 555–564
25. Wang L (2015) 3D-printed Solar Bytes Pavilion pushes the boundaries of architectural innovation. http://inhabitat.com/3d-printed-solar-bytes-pavilion-pushes-the-boundaries-of-architecturalinnovation/new-post-submission-icon-1039/. Accessed 20 June 2016
26. Weissenböck R (2014) Exploring the combination of laser cutting and robotic thermoforming as a technique for architectural envelopes. In: Mcgee W, Ponce de Leon M (eds) Robotic Fabrication in Architecture, Art and Design 2014. Springer, New York, pp 249–259
27. Yuan P, Meng H, Devadass P (2014) Performative tectonics—robotic fabrication methodology towards complexity. In: Mcgee W, Ponce de Leon M (eds) Robotic Fabrication in Architecture, Art and Design 2014. Springer, New York, pp 181–195

Chapter 5
Performative Architecture and Fiber Materials. Overview of the Main European Research Paths

Abstract A new field of exploration in the European research context on Informed Architecture is related to the digital innovations and fibrous materials to design and fabrication of units and technological systems as well as architectural organisms at 1:1 scale. To analyze this path, the contribution proposes a series of 1:1 scale pavilions as results of the applied research yearly conducted at the Institute for Computational Design Stuttgart (ICD) on the topic of Fibrous Tectonics. The research pavilion selected outlines the topics of the investigation: the importance of biology and the study of complex biological systems; the use of the material as an active agent capable to participate in the generation of form and structures through its physical characteristics; the adaptive manufacturing method is able to work in relation to the digital model and to expand the morphogenetic space of possible design solutions. The contribution allows gathering pros and cons in the three different investigative macroareas: performance-based design, material culture, and fabrication process. This analytical examination helps to create a clear research scenario around the concept as well as the definition of an innovative pathway for future studies, projecting into the future the assimilation of these innovative concepts in the building construction sector.

Keywords Composites systems · Parametric design · Fabrication strategies · Biology and architecture · Cyber-Physical making · Informed fibers structures

5.1 Research Lines on Composites Materials

A further research path on performative architectures and digital novelties concerns composite materials especially carbon and glass fibers. In this regard, the main source of inspiration is represented by fibrous systems, omnipresent in the biological world: studying the principles that regulate their composition and their evolution allows to abstract rules to be transferred in architecture through digital computation and innovative manufacturing processes. In fact, natural fibers are made up of a limited number of materials defined as constituents while the types of fibrous polymers present in nature are substantially four: cellulose in plants, collagen in animals, chitin in crustaceans and insects, and silk for arachnids [1].

A. Figliola and A. Battisti, *Post-industrial Robotics*,
https://doi.org/10.1007/978-981-15-5278-6_5

The ability to perform a wide range of functions is given by the following:

- composition of the elements;
- local variation of their density;
- orientation of the fibers;
- chemical composition of the binder, called matrix or reinforcement through which the tensile strength is increased.

The differentiation of the elements is due to the need to adapt and locally respond to specific performance parameters, to the distribution of forces as well as environmental conditions. Compared to the characteristics of natural fibrous systems, the excellent mechanical properties and their responsiveness represent the potential to be explored and then transferred to architecture through a process of abstraction based on scientific rules. The artificial threadlike materials, such as glass and carbon fibers, have the same features of natural fibrous systems: a high mechanical resistance, especially to tensile stress, low density, resistance to chemical agents, and atmospheric agents as well as good fireproof qualities. The glass or carbon fibers and polymer matrices in combination with traditional substances, such as plastic or concrete, gives rise to the composite materials used in architecture for the construction of fiber reinforced panels and structural elements as well as to produce technological networks for the automotive, naval, and aeronautical sectors. The compound materials present a series of application advantages such as lightness and excellent mechanical properties and some limitations that do not allow the complete affirmation in the AEC sector. One of the limitations in the creation of free form components derives from the intricate fabrication process which is expensive in terms of production times and costs. The commonly utilized methodologies are resin injection molding (RTM) and lamination: both procedures involve the creation of complex molds on which, after a series of operations, the composite is deposited. Further disadvantages in the use of compound materials are represented by the low resistance compared to concentrated structural loads, the poor surface quality as well as the difficulties deriving from the waste disposal. This technique was widely utilized at the end of the '80 s with the affirmation of computation and digital manufacturing technologies through which free-form molds are produced with a direct connection between digital prototype and CNC machine thanks to the utilization of numeric codes defined as g-code. In summary, composites materials in architecture provide for the combination of fibers with an additional material to increase their physical and mechanical properties and the use of molds for forming the component. This construction substance has been investigated by Greg Lynn both in university courses and for the realization of performance boats. The matter is discussed in the Greg Lynn and Foster Gage text of 2011 [2], "Composites, Surfaces, and Software: High Performance Architecture" in which the design of a performative boat realized by Yale University students is presented.

5.2 ICD, Institute for Computational Design: Fibrous Tectonics

The applied research conducted at the ICD presents an approach linked to the theories of morphogenetics design [3, 4] and emerging systems for what concerns the formal generation in relation to innovative production methods such as robotic manufacturing. Through the pavilions that summarize the studies carried out annually by the ICD in collaboration with the ITKE, it is possible to analyze the advances in research around the specific topic of *Fibrous Tectonics* [5]. The studies that have followed over the years have affected the development of material systems whose formal creation is a direct consequence of the relationship between material and manufacturing method, within a logic defined by Achim Menges as *Machine Computation*, constantly evolving compared to the technological development [6]. The experiments involved the generation of complex and optimized morphologies originated utilizing traditional substances, such as wood, and innovative elements such as fibers and composites exploiting in both cases the properties of the materials themselves as active agents. The combination of the computational and digital fabrication processes becomes the means by which to investigate new ways of optimally using materials commonly employed in the AEC sector, exploiting their physical and mechanical characteristics [7]. A decisive role is played by the interdisciplinary approach to research with the collaboration of other professionals concerning study of biological phenomena and their transfer in architecture, structural behavior, and engineering of non-linear and complex construction and, finally, development of innovative materials coupled with advanced production procedures. The research protocol outlines a series of cornerstones that can be found in almost all the studies conducted: the importance of biology and the investigation of complex biological systems to extract the structural and mechanical principles to be transferred in the design process; the importance of the material as an active agent able to participate in the form and structure creation through its physical characteristics; the adaptive manufacturing method able to work in relation to the digital model and to expand the morphogenetic space of possible design solutions [8]. Some fundamental properties of biological apparatuses are transmitted to the architectural project through the computational process and bottom-up approach [9]: heterogeneity, as a capacity for responsiveness of morphology with regard to many geometric parameters and external dynamic conditions; anisotropy, to define the structural arrangement in relation to the distribution of mechanical stress; hierarchy, as a non-linear manner of component composition. The understanding of biological systems is followed by the abstraction of fundamental concepts and then continues with a technological implementation to promote an architectural transfer. Of fundamental importance as a connection between the digital model and the physical prototype is a digital and robotic fabrication process capable of managing the complexity deriving from the introduction of biological principles in the design procedure and overcoming the geometric and morphological restrictions imposed by the industrial production operation leading to the massive reproduction of the same constructive element [10].

Using robots not as a simple tool to materialize a virtual model but as an active agent allows to expand the possibilities offered by the morphogenetic process. The terms Morphospaces of Robotic Fabrication and Machining Morphospaces [7] indicate the parametric connection between the morphological data determining the global shape and the conditions defining the robot setup. This associative relationship introduces a new space of design potentialities. Another research subject is represented by the investigation of the traits and assets of the materials capable of transforming into design opportunities to delineate informed morphologies. As stated by Achim Menges in the article "*Material Resourcefullness*" [9]: "The addition process to the ecological incentive, the computational process, therefore, the computation to explore its materially inherent design opportunities" the computational procedure becomes the means by which to discover the opportunities offered by the substance in relation to its complex and non-linear behavior. Starting from these premises, research projects have been developed to explore the structural and formal potential offered by composite material systems in relation to the computational process and innovative production methods.

5.3 Fibers Informed Structures: Case Studies

Although composites can be considered light materials, their application in architectural practice is linked to heavy components such as concrete or steel which use fibers to increase their physical and mechanical performance. In relation to the issue, the Fibrous Tectonics research path of the ICD and ITKE Stuttgart experiments innovative processes in the construction of a series of pavilions at the 1:1 scale aimed at demonstrating how the combination of biomimetics, digital computation, and robotic fabrication makes it feasible to transfer the abstract biological principles in the common architectural practice. Due to digital innovations, it is possible to expand the applications of compound materials in the architectural and structural field as well as to experiment with new production operations.

The main points of the research conducted at ICD jointly with ITKE are as follows:

- study of biological fibrous systems and definition of the principles that regulate their structure and their mechanical properties;
- experimentation of production processes able to limit the use of expensive formworks, to reduce waste material, maintaining a high degree of geometric freedom;
- definition of physical and mechanical properties of fibers and binders in relation to the manufacturing procedure adopted.

To analyze the different approaches, the contribution proposes a series of case studies as results of three applied researches, the 1:1 scale pavilion realized by the ICD-ITKE students (Fig. 5.1).

The case studies are selected verifying the correspondence to the following parameters:

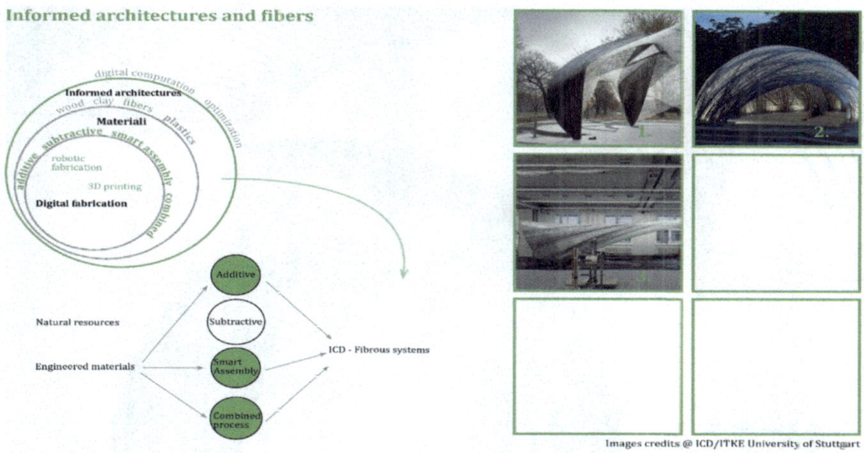

Fig. 5.1 Informed architectures and fibers. Overview of the main research topics and case studies. © Angelo Figliola

- presence of a performance-based process through which explore informed architectures;
- utilization of fibers material, such as glass fibers and carbon fibers;
- materialization of the digital model through innovative manufacturing procedures, specifically robotic fabrication.

5.3.1 ICD/ITKE Research Pavilion 2012

The *Research Pavilion 2012/2013*[1] [6, 11] (Fig. 5.2) realized by the ICD Stuttgart is the first of a series of architectural prototypes resulting from research on the application of composite materials for the construction of informed architectures through advanced modeling techniques and robotic fabrication. As will be mentioned below, the study conducted by ICD is based on the analysis of biological organisms and the consequent abstraction of physical and mechanical principles to apply in the scientific process of morphology generation.

The concept from which the Research Pavilion 2012 was generated, arises from the observation and study of the exoskeleton of lobsters (Homarus Americanus) characterized by a material differentiation, a softer and a harder portion, in relation the functional and structural performances. The examined biological apparatus is characterized by a strong material heterogeneity as well as a modification of the fiber's orientation of which it is composed; these properties make it possible to respond

[1]Institute for Computational Design—Prof. Achim Menges, Institute of Building Structures and Structural Design—Prof. Jan Knippers. Full credits at: http://icd.uni-stuttgart.de/?p=8807.

Fig. 5.2 ICD/ITKE *Research Pavilion 2012*. © ICD/ITKE University of Stuttgart

differently to local variations related to performance parameters as well as to optimize substance distribution along the surfaces. The design strategy was created from the abstraction of the biological principles described above: the structural analysis allowed to alter the fibers in response to local variations in the structural stress, the use of two distinct types of composites (glass and carbon fibers) has allowed to minimize the utilization of mold to facilitate the winding of the compounds while the alteration of the orientation of the fibers has permitted the generation of an edifice able to control in an adaptive way the distribution of loads. The construction of the structure with an 8 m diameter and with a 3.5 m height was possible due to the use of a steel mold mounted on a rotating table, capable of moving around 360°, on which the fibers are wound utilizing a robot arm placed in a high position with regard to the ground floor (Fig. 5.3). During the process two kinds of fibers were used: glass fibers for the spatial articulation; carbon fibers for structural purpose, and uniformly distribute loads along the surface. This experimentation enables the exploitation of additional type of filaments useful for monitoring the structural performance thanks to the possibility of integration into the structure itself. The research pavilion created by the ICD represents the starting point of a research path that studies the application of composite materials in the architectural and engineering fields through robotic manufacturing processes with reference to biomimetic principles. The aims underlying the research are manifold: starting from the simplification of the fabrication

Fig. 5.3 ICD/ITKE *Research Pavilion 2012*: robotic fabrication process of fibers winding. © ICD/ITKE University of Stuttgart

operations of compounds made with composite elements, traditionally utilized in the naval and aerospace sectors, up to the realization of complex structures able to benefit from the properties of the material used. Within this scenario, digital computing and robotic fabrication play a fundamental role in achieving the objectives described above. The connection between the digital model and the analysis of the structural performances makes it possible to generate performative and informed morphologies and define a hierarchical and heterogeneous dislocation of the material as well as managing the orientation of the fibers and therefore controlling the distribution of the loads.

The transfer of complex morphologies, created in the digital space, in the architectural practice takes place through the use of anthropomorphic robotic arms and customized tools able to manage the elaborate correlation between digital and physical space. The utilization of the described design strategy (Fig. 5.4) consented to experiment with a new application for composite materials and to test their structural behavior. Critical issues remain regarding the management of the complex digital and production process as well as a limitation on the dimensional scalability of the components that can be realized through this design strategy.

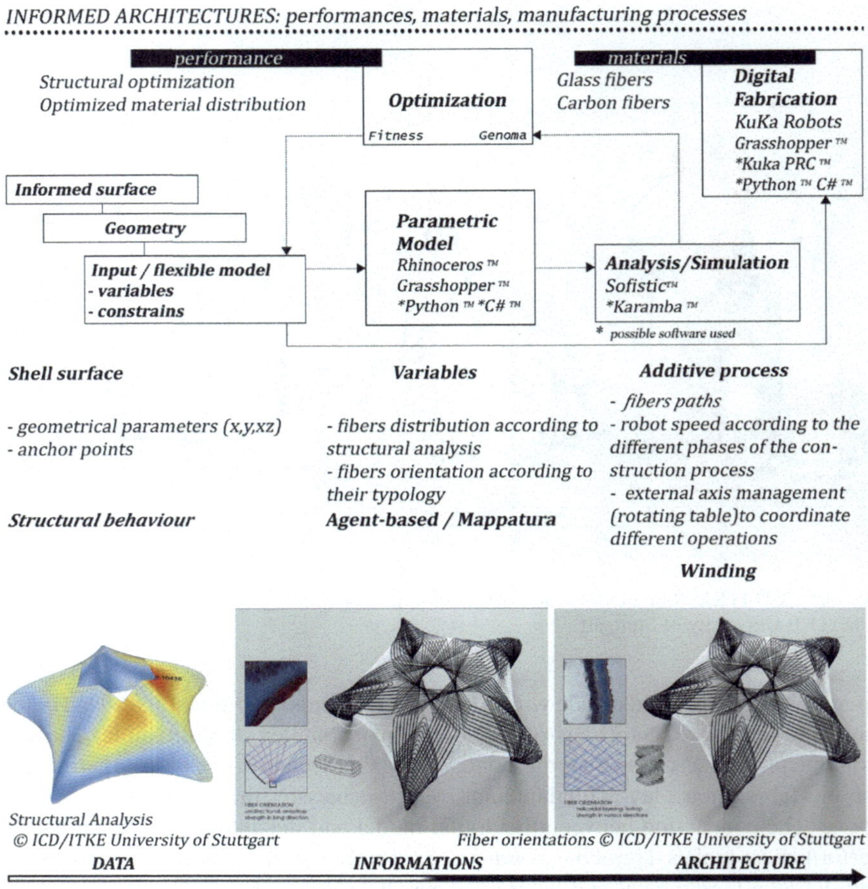

Fig. 5.4 ICD/ITKE University of Stuttgart, *Research Pavilion 2012*: performances, materials, manufacturing processes. A design methodology. © Angelo Figliola

5.3.2 ICD/ITKE Research Pavilion 2014–'15

The 2014 research pavilion[2] [12–14] (Fig. 5.5) produced by the ICD Stuttgart proposes a responsive morphology result of a computational and fabrication process informed by a series of parameters such as the characteristic of materials used, the structural concept, and the production procedure of robotic manufacturing. The proposed technological system starts from the study of a natural phenomenon, instead of a biological organism, the construction of the protective shell that allows the survival of an animal species, in this case the water spiders.

[2]Institute for Computational Design—Prof. Achim Menges, Institute of Building Structures and Structural Design—Prof. Jan Knippers. Full credits at: http://icd.uni-stuttgart.de/?p=12965.

Fig. 5.5 ICD/ITKE *Research Pavilion 2014–'15*. © ICD/ITKE University of Stuttgart

The arachnid builds a support network to the air bubble chosen in such a way as to create a protected environment in which to survive immersed in water. By studying the organization of the system and its training principles, the pavilion is conceived as a shell, inflatable structure, made of ETFE and subsequently reinforced by carbon fibers distributed through a robotic arm. The simulation of the structural behavior through FEA analysis and the consequent optimization of the construction in relation to loads such as wind, snow, and weight, determined the distribution of the fibers on the surface of ETFE. By adopting this methodology, the industrial robot becomes an active agent capable of reacting and adapting to the real conditions of manufacture as well as distributing the material only where necessary with consequent maximization of substance resources (Fig. 5.6).

Fig. 5.6 ICD/ITKE *Research Pavilion 2014–'15*: bio-inspired and adaptive robotic fabrication process. © ICD/ITKE University of Stuttgart

The creation of a feedback loop system between the digital model and the production process was possible thanks to the customization of the end effector and the introduction of proximity and position sensors able to correct the positioning of the robot with regard to variable fabrication conditions, for example, pressure of oscillation, flexion of the membrane, and speed of the robot. The result of the proposed methodology is an informed architectural organism of 40 m^2 with a weight of 260 kg, capable of adapting the structure and production procedure to external inputs acquired in real-time, generating a connection between the digital and physical model. The research pavilion built at the ICD in Stuttgart is the result of a complete design process capable to associate the digital and physical model through a feedback loop between simulation and prototype. The computational procedure was informed by the structural performance and optimized for the distribution of the material in an ETFE surface system, reinforced through carbon fibers, and at the same time for the structure and envelope. The robotic manufacturing process informs the project by defining variables and constraints and consequently creating a space of building line possibilities together with the meta-project. The robotic arm employment for the distribution of the fibers along the lines of structural stress binds the dimensions of the artefact to the dimensional parameters that characterize the radius of action of the industrial robot (Fig. 5.7). The dimensional aspect of the robot workspace represents a criticality that can turn into an opportunity for the organization of the constructive and technological network in the various phases of the project. The informed architectural organism (Fig. 5.8) has some features that make it scalable for architectural applications, both as a structural element and for multi-layer architectural envelopes. The operation of material maximization according to structural performance, the weight of the building and the efficiency of the production process, characterize the 1:1 prototype and express its potential. Finally, the replacement of carbon fibers with ecological composites (biocomposites) represents a further realizable implementation of the procedure.

5.3.3 ICD/ITKE Research Pavilion 2016–'17

The 2016–2017 research pavilion[3] [15–17] (Fig. 5.9) made by the ICD Stuttgart is presented as a continuation of the research path on the use of composite materials for the fabrication of complex structures based on the study of biological principles. The functional concepts and the logic that govern the construction process are abstracted from the observation of two biological organisms, the Larvae Lyonetia Clerkella and Leucoptera Erythrinella, during the realization of the filamentous silk structure through which they colonize the leaves.

The bottom-up approach has allowed to conceptualize some biological procedures to define the design and construction strategy starting from the fibers winding

[3]Institute for Computational Design—Prof. Achim Menges, Institute of Building Structures and Structural Design—Prof. Jan Knippers. Full credits at: http://icd.uni-stuttgart.de/?p=18905.

Fig. 5.7 ICD/ITKE *Research Pavilion 2014/15*: cyber-physical making. © ICD/ITKE University of Stuttgart

methods, in order to refrain from the use of molds and formworks, up to the hierarchy and the orientation of the fibers in relation to the structural complexity and the constructional dimensions. The design of the pavilion starts from the determination of the shell surface and from the structural analysis necessary to establish the distribution of stress and the consequent placement of the filaments according to a hierarchy determined by the result of the analysis conducted. The impressive dimensions of the structure, 12 m span and only two supports, together with the desire to avoid the utilization of expensive formwork to deposit the fibers, required the development of

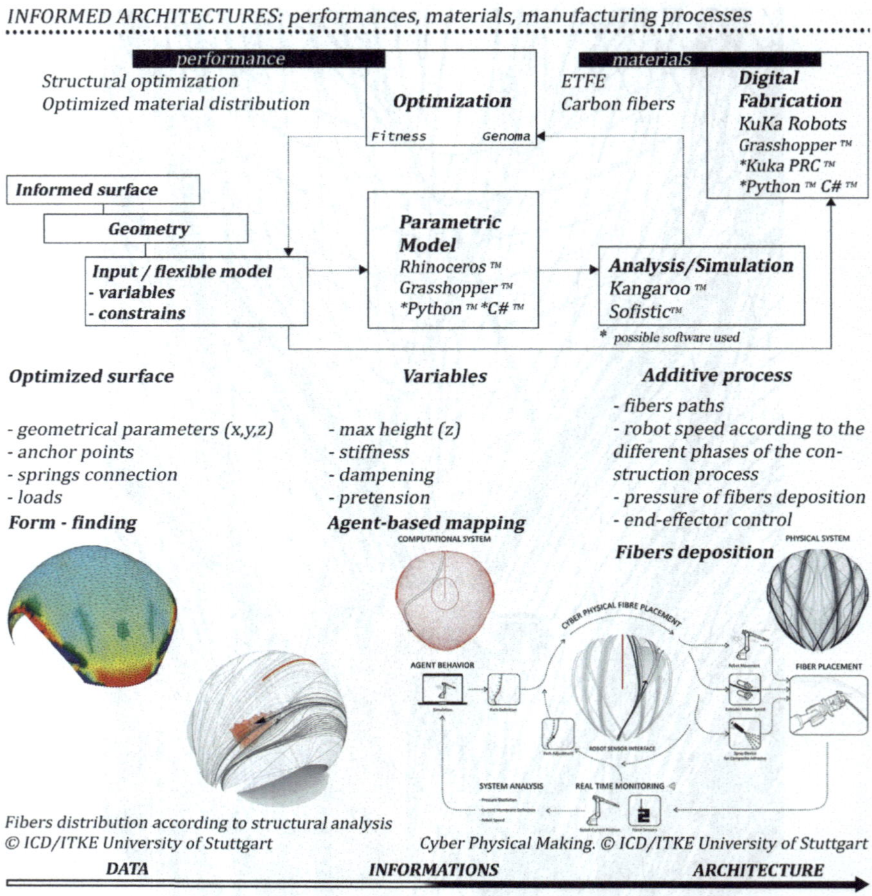

Fig. 5.8 ICD/ITKE University of Suttgart, *Research Pavilion 2014/15*: performances, materials, manufacturing processes. A design methodology. © Angelo Figliola

a complex manufacturing strategy based on the cooperation between two kinds of autonomous machines such as industrial robots and drones (Fig. 5.10).

The limitations described above deriving from the dimensions of the robots are overcome thanks to the use of a drone, appropriately designed and programmed, through which the continuity in the deposition of the fibers is guaranteed by the two robotic arms placed at the ends of the structure. The correct development of the production process is ensured by the presence of sensors that enable communication between the physical and the digital space in such a way as to continuously adapt the two procedures and manage the cooperation between the various tools involved. The result of the experimentation conducted at the ICD in Stuttgart, after the pavilions of 2013–2014 and 2015–2016, represents a further step forward compared to research

Fig. 5.9 ICD/ITKE *Research Pavilion 2016/17*. © ICD/ITKE University of Stuttgart

Fig. 5.10 ICD/ITKE *Research Pavilion 2016/17*: cooperation between robots and drones for winding the fibers. © ICD/ITKE University of Stuttgart

on the utilization of composite materials for the construction of complex structures and collaborative fabrication of cyber-physical making (Fig. 5.11). The construction of the pavilion made it possible to verify the previously used workflow and make implementations with regard to the digital manufacturing operation. The analysis of the structural performance of the shell surface determined the hierarchy and the distribution of the two types of fibers in relation to stress values. To overcome the

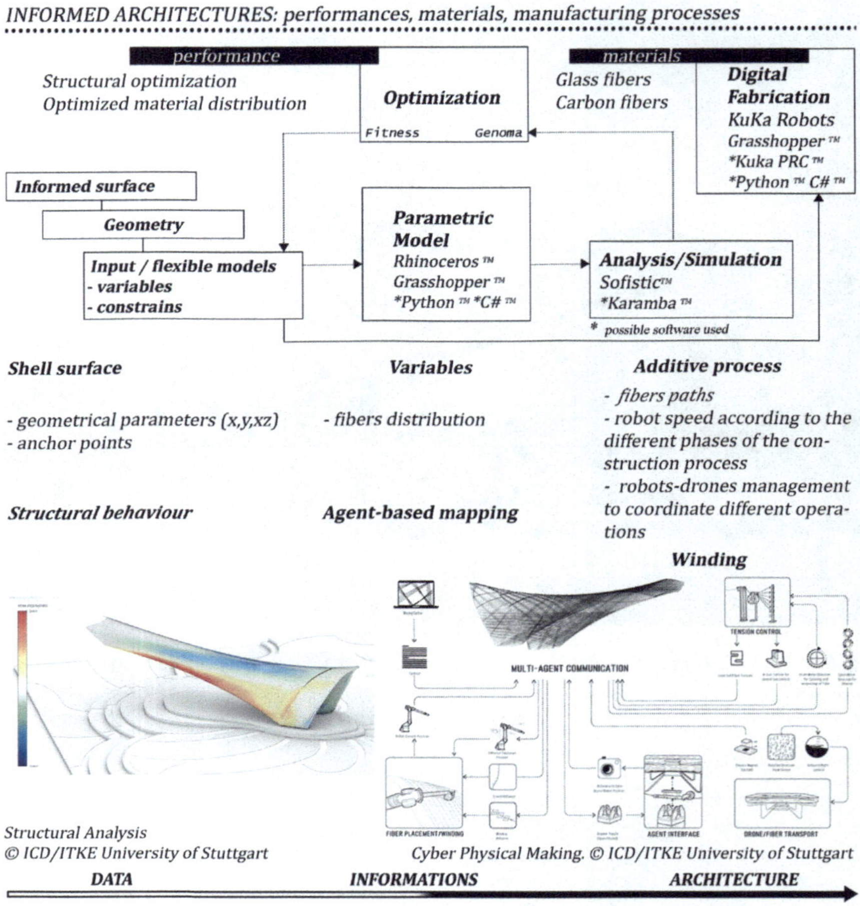

Fig. 5.11 ICD/ITKE University of Suttgart, *Research Pavilion 2016/17*: performances, materials, manufacturing processes. A design methodology. © Angelo Figliola

dimensional limits that characterize the production with industrial robots, reachability, a further tool is introduced, a drone, able to assist the two robots and extend the work area. The cooperation between the various devices allows to overcome one of the limits highlighted by previous project experiences. The relationship system is managed by a complex and sophisticated sensor system capable to coordinate the action of the different tools present in the workspace. Although the management of the production process is complicated and difficult, the 2016–2017 pavilion represents an important step for a potential introduction into the AEC sector of the proposed design strategies. The technological system was developed in the laboratory and subsequently transported to the installation site. The procedure opens new possibilities in industrial prefabrication for architectural and structural applications.

5.4 Conclusion: Informed Composites Systems

The research developed at the ICD through the realization of prototypes on a 1:1 scale showed how it is possible to transfer biological principles that regulate the functioning of natural systems in order to realize optimized buildings using composite materials, fibers, and resins. Thanks to digital computation it is possible to maximize the morphology informing the design process and including the parameters related to the material used and the production procedure adopted. To generate complex structures characterized by non-linear behavior, it was necessary to develop innovative constructive methodologies that employ robots and customized end effectors in relation to a particular fabrication process. Thus, the research conducted at ICD has introduced the Cyber-Physical Making concept: the machine becomes an extension of the artisan designer who perceived and analyzed the physical properties of a material through his own hands [18]. This methodological approach is a design response to the theory of Soft Systems or non-linear, flexible, adaptable, and continuously evolving systems whose dynamism is constantly fed by a flow of information coming from outside [19]. The ability to act and react in a dynamic manner, presenting a certain quotient of intelligence, makes the system sensitive to variations due to the interaction between the parties involved. Concerning the research application fields, the focus is placed on the structural scope, given the excellent performance of the material system and its reduced weight when compared with traditional structural systems. Within the research evolution, it was possible to move from the assembly of optimized technological units realized off-site, to the definition of manufacturing protocols that benefit from the collaboration between different types of autonomous machines, robots, and drones, able to cooperate to produce the structural component directly on-site. This methodology has permitted to overcome the critical issues linked to the dimensional aspects of the instruments utilized, opening new research scenarios for the construction of structural technological systems as well as multi-layer building envelopes. With regard to technology transfer in the AEC sector, critical issues persist regarding the complexity of the design and construction process in the maximization of the morphology, the structural discretization, and the setup of the robotic fabrication. Further implementation focuses on developing innovative material systems based on biocomposite and naturally derived resins and on tech transfer process, from academic research to AEC sector to challenging the future urban development. Regarding the material system, polymeric composites reinforced with natural fibers represent a possible alternative to traditional composites both for the renewable origin and for the affordability of the whole operation.

ICD/ITKE University of Stuttgart Projects Credits ICD/ITKE Research Pavilion 2012: ICD Institute for Computational Design—Prof. Achim Menges; ITKE Institute of Building Structures and Structural Design—Prof. Jan Knippers; Concept Development: Jakob Weigele, Manuel Schloz; System Development & Realization: Sarah Haase, Markus Mittner, Josephine Ross, Manuel Schloz, Jonas Unger, Simone Vielhuber, Franziska Weidemann, Jakob Weigele, Natthida Wiwatwicha with the support of Michael Preisack, Michael Tondera (Faculty of Architecture Workshop); Scientific Development & Project Management: Riccardo La Magna (structural design), Steffen Reichert (detail design), Tobias Schwinn (robotic fabrication), Frédéric Waimer (fibre composite technology

& structural design); In collaboration with University of Tuebingen, Departement of Evolutionary Biology of Invertebrates—Prof. Oliver Betz, Departement of Palaeontology of Invertebrates—Prof. James Nebelsick ITV Denkendorf—Dr.-Ing. Markus Milwich; Funding: KUKA Roboter GmbH, Competence Network Biomimetics, SGL Group Momentive, AFBW—Allianz Faserbasierte Werkstoffe Baden-Württemberg, FBGS Technologies GmbH, MFTech SARL, Minda Schenk Plastic Solutions GmbH, Stiftungen LBBW, Südwestbank AG, Wayss & Freytag Ingenieurbau AG.

ICD/ITKE Research Pavilion 2014–15: ICD Institute for Computational Design—Prof. Achim Menges; ITKE Institute of Building Structures and Structural Design—Prof. Jan Knippers; Scientific Development: Moritz Dörstelmann, Valentin Koslowski, Marshall Prado, Gundula Schieber, Lauren Vasey; System Development, Fabrication & Construction: WS13/14, SoSe14, WS14/15: Hassan Abbasi, Yassmin Al-Khasawneh, Yuliya Baranovskaya, Marta Besalu, Giulio Brugnaro, Elena Chiridnik, Tobias Grun, Mark Hageman, Matthias Helmreich, Julian Höll, Jessica Jorge, Yohei Kanzaki, Shim Karmin, Georgi Kazlachev, Vangel Kukov, David Leon, Kantaro Makanae, Amanda Moore, Paul Poinet, Emily Scoones, Djordje Stanojevic, Andrei Stoiculescu, Kenryo Takahashi and Maria Yablonina; WS14/15: Rebecca Jaroszewski, Yavar Khonsari, Ondrej Kyjanek, Alberto Lago, Kuan-Ting Lai, Luigi Olivieri, Guiseppe Pultrone, Annie Scherer, Raquel Silva, Shota Tsikoliya. With the support of: Ehsan Baharlou, Benjamin Felbrich, Manfred Hammer, Axel Körner, Anja Mader, Michael Preisack, Seiichi Suzuki, Michael Tondera. In collaboration with: Departement of Evolutionary Biology of Invertebrates, University of Tuebingen Prof. Dr.Oliver Betz, Departement of Palaeontology of Invertebrates, University of Tuebingen, Prof. Dr.James Nebelsick, Dr.Christoph Allgaier Institute for Machine Tools, University of Stuttgart Dr. Thomas Stehle, Rolf Bauer, Michael Reichersdörfer Institute of Aircraft Design, University of Stuttgart, Stefan Carosella, Prof. Dr.-Ing. Peter Middendorf. Funding: KUKA Roboter GmbH, GETTYLAB, tat aiRstructures, SGL Carbon SE, Sika Deutschland GmbH, Daimler AG, Walther Spritz- und Lackiersysteme GmbH, Lange + Ritter GmbH, Gibbons Fan Products Ltd, igus® GmbH, Peri GmbH, HERZOG Maschinenfabrik GmbH & Co. KG, AFBW—Allianz Faserbasierter Werkstoffe Baden-Württemberg e.V., Reinhausen Plasma GmbH, Reka Klebetechnik GmbH, HECO-Schrauben GmbH & Co. KG, Airtech Europe S.A., Mack Gerüsttechnik GmbH, RentES, Stahlbau Wendeler GmbH + Co. KG, CARU Containers GmbH, EmmeShop Electronics, STILL GmbH, SH-Elektrotechnik, GEMCO, Zeppelin Rental GmbH & Co. KG.

ICD/ITKE Research Pavilion 2016–17: Project Team ICD Institute for Computational Design and Construction—Prof. Achim Menges; ITKE Institute of Building Structures and Structural Design—Prof. Jan Knippers; Scientific Development: Benjamin Felbrich, Nikolas Früh, Marshall Prado, Daniel Reist, Sam Saffarian, James Solly, Lauren Vasey; System Development, Fabrication and Construction: Miguel Aflalo, Bahar Al Bahar, Lotte Aldinger, Chris Arias, Léonard Balas, Jingcheng Chen, Federico Forestiero, Dominga Garufi, Pedro Giachini, Kyriaki Goti, Sachin Gupta, Olga Kalina, Shir Katz, Bruno Knychalla, Shamil Lallani, Patricio Lara, Ayoub Lharchi, Dongyuan Liu, Yencheng Lu, Georgia Margariti, Alexandre Mballa, Behrooz Tahanzadeh, Hans Jakob Wagner, Benedikt Wannemacher, Nikolaos Xenos, Andre Zolnerkevic, Paula Baptista, Kevin Croneigh, Tatsunori Shibuya, Nicoló Temperi, Manon Uhlen, Li Wenhan. With the support of Michael Preisack and Artyom Maxim. In collaboration with: Institute of Aircraft Design (IFB)—Prof. Dr.-Ing. P. Middendorf, Markus Blandl, Florian Gnädinger, Institute of Engineering Geodesy (IIGS)—Prof. Dr.-Ing. habil. Volker Schwieger, Otto Lerke, Department of Evolutionary Biology of Invertebrates, University of Tuebingen—Prof. Oliver Betz Department of Palaeontology of Invertebrates, University of Tuebingen—Prof. James Nebelsick. Supported by Volkswagen Stiftung, GETTYLAB, Kuka Roboter GmbH, Peri GmbH, SGL Technologies GmbH, Hexion Stuttgart GmbH, Ed. Züblin AG, Lange Ritter GmbH, Stahlbau Wendeler GmbH, Leica Geosystems GmbH, KOFI GmbH.

References

1. Menges A, Knippers J (2015) Fibrous tectonics. Menges, a material synthesis: fusing the physical and the computational. AD architectural design. John Wiley & Sons, London, pp 40–47
2. Lynn G, Gage MF (2011) Composites, surfaces, and software: high performance architecture. Actar, Yale School of Architecture
3. Hensel M, Menges A (2006) Morpho-ecologies. AA Publications, London
4. Michael H, Achim M (2008) Designing morpho-ecologies: versatility and vicissitude of heterogeneous space. Archit Des 78(2):102–111
5. Menges A (2015) Material synthesis. AD architectural design. John Wiley & Sons, London
6. Menges A (2015) The new cyber-physical making in architecture. A material synthesis: fusing the physical and the computational. AD Archit. Des 237:40–47
7. Menges A (2012) Morphospaces of robotic fabrication. In: Brell-Çokcan S, Braumann J (eds) Robotic fabrication in architecture, art, and design. Springer, Wien, New York, pp 28–47
8. Knippers J, Gabler M, La Magna R, Waimer F, Menges A, Reichert S, Schwinn T (2014) From nature to fabrication: biomimetic design. Principles for the production of complex spatial structures. In: Hesselgren L, Sharma S, Wallner J, Baldassini N, Bompas P, Raynaud J (eds) Advances in architectural geometry 2014. Springer, Wien NewYork, pp 91–106
9. Menges A (2012) Material resourcefulness: activating material information in architectural design. Material computation: higher integration in morphogenetic design. AD Archit Des 216:34–44
10. Krieg OD, Dierichs K, Reichert S, Menges A (2011) Performative architectural morphology finger-joint plate structures integrating robotic manufacturing, biological principles and location-specific requirements. In: Gengnagel C, Kilian A, Palz N, Scheurer F (eds) Proceedings of computational design modelling. Springer, pp 259–266
11. ICD Stuttgart (2012) ICD Research Buildings/ Prototypes. http://icd.uni-stuttgart.de/?p=8807 Accessed 12 Apr 2017
12. Doerstelmann M, Knippers J, Koslowski V, Menges A. Prado M, Schieber G, Vasey L (2015) ICD/ITKE Research pavilion 2014–15–fibre placement on a pneumatic body based on a water spider web. In: Menges A (eds) Material Synthesis: Fusing the physical and the computational. AD architectural design, vol 237, pp 60–65
13. Schieber G, Koslowski V, Dörstelmann M, Prado M, Vasey L, Knippers J, Menges A (2015) Integrated design and fabrication strategies for fibrous structures. In: Thomsen R, Tamke M, Gengnagel M, Faircloth C, Scheurer F (eds) Modelling behaviour proceedings of the design modelling symposium Copenhagen 2015. Springer, Cham Heidelberg New York Dordrecht London, pp 237–246
14. Vasey L, Baharlou E, Dörstelmann M, Koslowski V, Prado M, Schieber G, Menges A, Knippers J (2015) Behavioral design and adaptive robotic fabrication of a fiber composite compression shell with pneumatic formwork. In: Combs L, Perry C (eds) Computational ecologies: Design in the Anthropocene, Proceedings of the 35th Annual conference of the association for computer aided design in architecture. University of Cincinnati, Cincinnati OH, pp 297–309
15. Felbrich B, Früh N, Prado M, Saffarian S, Solly J, Vasey L, Knippers J, Menges A (2017) Multi-Machine fabrication: an integrative design process utilising an autonomous UAV and industrial robots for the fabrication of long-span composite structures. In: Nagakura T, Tibbits S, Mueller C (eds) ACADIA–Disciplines & disruption proceedings of the ACADIA Conference 2017. Acadia Publishing Company, Cambridge, MA, pp 248–259
16. ICD Stuttgart (2017) ICD Research Buildings/ Prototypes. http://icd.uni-stuttgart.de/?p=12965 Accessed 12 Apr 2017
17. Solly J, Frueh N, Saffarian S, Prado M, Vasey L, Felbrich B, Reist D, Knippers J, Menges A (2018) ICD/ITKE Research pavilion 2016/2017: integrative design of a composite Lattice Cantilever. In: Mueller C, Adriaenssens S (eds) creativity in structural design, proceedings of the IASS Conference 2018. International Association for Shell and Spatial Structures (IASS), Boston, pp 1–8
18. Carpo M (2014) Mario Carpo in conversation with Matthias Kohler. In: Gramazio F, Kohler M, Langenberg S (eds), Fabricate: Negotiating Design & Making. Gta-Verl, Zurich, pp 12–21
19. Kwinter S (1993) Soft system. In: Culture Lab. Princenton Architectural Press, pp 206–227

Chapter 6
Feedback on the Design Processes for the Materialization of Informed Architectures

Abstract The analysis of case studies and applied research leads to the definition of a design methodology for computing and fabricating informed architectures, stimulating a critical reflection on the possibility of introducing the technological systems proposed in architectural practice, explaining the level of applicability of the same and identifying the main critical aspects of the processes. Such procedures are the necessary basis for delineating innovative pathways for future studies. In order to state that technology is the answer to problems related to design processes, citing the words of Cedric Price in 1966 ("*Technology is the answer but what was the question?*" Cedric Price, 1966), it is essential to first define the matter in a clear way, to ensure that the contribution is not compromised, losing its relevance, due to a generalist thought that sees the use of emerging technologies as a solution to all critical design issues. Once the methodology that sets a design and construction operation paradigm shift has been defined, the next step of the research concerns the concrete possibility of a technological transfer from the academic world to the actors involved in the design and production process, including the industries in the sector. The following questions may arise. Among the technological networks studied, which ones can be introduced in architectural-executive practice? What are the critical issues that prevent a concrete application outside the experimental field of academic research?

Keywords Design methodology · Digital computing · Innovative manufacturing techniques · Informed architectures · Technological system · Construction processes

6.1 Feedback from the Analysis of the Case Studies

Through the examination of case studies, it was possible to verify how parametric models' flexibility allows to include performative parameters, as well as information related to the manufacturing process, in the early stages of the design procedure. The correlation between material and fabrication, theorized by Oxman [10] as "MFD: Material-Fabrication-Design", is determined thanks to the computational

A. Figliola and A. Battisti, *Post-industrial Robotics*,
https://doi.org/10.1007/978-981-15-5278-6_6

process' information which increases the tectonic relations between structure and material within the limits of the digital fabrication logics adopted. The analysis of the projects highlighted the importance of two tools of the methodology: the construction of the meta-project to define the variable and invariable design criteria, and the information of the process through the data-driven strategy. The flexibility of the models and the data of the operations allows to visualize and investigate a series of design options aimed at identifying the optimal solution, or at the minimum the best solution, that responds to the parameters indicated by the designer through the variation of the meta-project. Through the new tectonics,[1] resulting from the application of the methodology briefly described above, it is possible to establish the concept of informed architecture[2] able to optimize performance criteria through specific morphological configurations, giving life to a visually static idea of responsiveness rather than a visually dynamic one [7]. The computational process links the versatile geometric parameters to the results of the performance analyzes and includes in a single operation the design criteria, the performance evaluation, and the parameters of the manufacturing procedure. If we see the flexibility of the parametric models as one of the potentials of the methodology, the increase in its variables is one of its crucial points. Critical issues can be considered directly proportional to the complexity of the data structure describing the relationships between geometric criteria and evaluation of performances (conducted in the early stage design phase). The complexity of parametric models grows alongside genetic optimization processes and the transfer of procedures to the 1:1 scale for architectural and structural applications. In the first case, the use of genetic maximization operations involves the use of elaborate algorithms whose management requires advanced skills in data structuring, while the transfer of digital processes to the 1:1 scale requires an advanced control of all phases of the project, from the creative aspects to construction details. The parametric model represents the condensing of all the procedures, from geometry definition to the production, leading to significantly different management of the complexity compared to traditional processes. From the fabrication point of view, the analysis of the case studies allowed to define some critical features: the dimensions of the tools and the useful working area rather than the production volume of the robotic arms. Such characteristics represent a limit to the project strategies that can be adopted; most of the case studies propose technological systems and study spatial aggregation strategies enhancing the possibilities offered by the technologies. The absence of an aggregative strategy represents one of the problematic aspects in the components' application at the 1:1 scale, together with the dimensional limits of the fabrication area and the difficult management of the computational process. In this regard, the materialization of complex structures represents a further critical aspect as it requires the design of customized and highly sophisticated end effectors in its mechanical and

[1]The term tectonics refers to the syntax of the construction, i.e., how the individual pieces of an architectural structure are organized in regard to a constructive logic.

[2]The term architecture refers not only to the architectural organism but also to all the parts that compose it, as units or technological systems.

electronic parts, as well as the introduction of sensors and the design of feedback loop systems between the digital model and the prototype.

The analysis of the case studies highlighted the following facts:

- The information of the design process to obtain optimized and informed morphological configurations based on performance parameters used as input in the early stage design phase;
- The data-driven strategy and the construction of flexible models, through the meta-project, to link in a single workflow the geometric criteria describing the shape with simulation and execution analysis;
- The maximization operation, obtained through genetic algorithms or search algorithms,[3] as an important tool for exploring the association between geometric parameters and performance evaluation aimed at searching optimized solutions;
- The design-material-manufacturing correlation to strengthen the relationship between structure and material within the limits of the digital manufacturing logics utilized;
- The inclusion of functional parameters in the fabrication of elements to guarantee actual producibility, eliminating the fallacy that characterized the first digital age or the mismatch between the virtual form and the physical model;
- The investigated digital fabrication technologies to reintroduce vernacular materials with high-performance traits into the contemporary architectural debate;
- The explored digital manufacturing technologies propose additive and subtractive processes to place or remove the material only where necessary based on the information of the operation;
- The operational methodology, the relationship between design and fabrication, to introduce a creative process and exploration of innovative formal codes for informed architectures.

6.2 Design Methodology: Digital Computing and Innovative Manufacturing Techniques for the Exploration of Informed Architectures

One of the objectives of the projects' analysis is the proposal for a new methodology to design and fabricate informed architectures able to operate on the edge of different disciplinary areas, to synthesize the information collected and extract the main concepts. The proposed methodology introduces a series of design instruments to operate informing the project based on performance parameters selected as project inputs and able to vary following the mutations of the application context. The idea of informed architecture is presented as an alternative scenario in the sphere of performance-based architecture [1, 4–6, 8, 9] reinforcing the notion of visually static responsive architecture. Beyond the theoretical aspects discussed in the previous

[3] Search Algorithms use computing power to explore different design solutions aimed at visualizing their effects on performance parameters chosen as project objectives.

chapters, it is important to underline the instrumental contributions of the methodology which allows to integrate the performative parameters in the early stage of the project and to include details related to manufacturing processes. The meta-project represents one of the tools and is designed through flexible parametric models to define the variable and invariable geometric criteria and to link them to the simulation and evaluation of the performance parameters. In the sector of process information, the analysis introduces the concept of data-driven design [11] as an instrument through which qualitative and quantitative data become guiding parameters in order to make informed design decisions [3], an operation in which data becomes an absolute priority. Through this design tool, it is possible to eliminate the false positives assumptions, and preconceptions characterizing the traditional design process. As for other branches of application, even in architecture, the data represent nothing but numbers, raw material, which needs to be organized through the construction of informed models that allow the operator, in this case, the designer, to make relevant decisions in terms of performance optimization. The raw material is transformed through specific know-how into information and subsequently in useful and relevant knowledge to make a design choice or to make a specific decision within the space of design potentialities. In this regard, the possibility of integrating performance simulation with the formal generation process, connecting quantitative values with geometrical parameters, opens up computational and form searching optimization paths. Genetic maximization is one of the innovations introduced as an operative tool for the generation of informed architectures. The hierarchical and structured data at the base of the parametric models define the genotype of the architecture and links its variation to the phenotype, represented by the fitness values or quantitative parameters resulting from the analysis of performances. One of the results of the process described, which passes from the meta-project to the manufacturing procedure, is a space of design possibilities where the designer can choose the solution resulting from the integration of formal choices and action of external forces on them. By applying such methodology, it is possible to overcome the typological paradigm with a continuous formal variation according to the change of the boundary conditions of the system. The synthesis made for the definition of the methodological process introduces a further performative layer: the power of digital computation, morphological generation, is enriched by a material connotation thanks to the integration of the criteria underlining the production and the materialization of complexity in the early stage design phase of units, technological networks, and architectural organism. The design-material-fabrication correlation is defined due to the information of the computational procedure that increases the tectonic relations between structure and material within the limits of the digital manufacturing logics adopted. The relationship mentioned above is the starting point for exploring new design paradigms based on the concept of form-finding; that is, the design of a process, whose formal creation is guided by the achievement of pre-established performance parameters, rather than a final form.

The design paradigms for informed architectures can be summarized as follows:

- Variable geometric patterns in relation to the execution parameters that inform the design process;
- Tiling and population of complex surfaces through modules whose distribution is informed by performance, structural, and energy-environmental criteria;
- Informed distribution of the material based on the evaluation of stress lines, flow analysis, structural patterns, and XESO optimization [2];
- Euclidean geometric transformations [12].

This approach can be reported in a matrix in which the digital manufacturing technologies and the respective production methods are correlated with the material and the design paradigm that ought to be explored in order to achieve the desired performance parameters. A new era begins thanks to the integration between informed architecture and innovative digital fabrication technologies founded on the performative customization of architectural systems, from technological units to the architectural organisms. The hybrid space of iteration between designer and machine is contaminated by other disciplines, investigating manufacturing methods but also stimulating creativity through a fruitful process of collaboration. Interdisciplinarity opens up a post-industrial era in which aspects linked to analogical design and production methodologies are interrelated with disruptive digital technologies.

6.3 Tech Transfer: From Research to Architectural Practice

The examination of case studies and employed research lead to a critical reflection on the possibility of introducing the technological systems proposed in architectural practice, exploring the level of applicability of the latter and identifying the main critical aspects of the processes, fundamental steps for delineating innovative future research pathways. Among the technological systems studied, which ones can be introduced in the architectural field? What are the critical issues that prevent a concrete application outside the experimental field of academic research? The first consideration for a correct reading of the application context concerns the continuous interference and contamination between the technical elements buildings and environmental systems. In this regard, some of the systems analyzed present technological solutions that go beyond the definition of common classes for technological units and technical elements characterizing the architectural context. A synthesis is necessary to outline the area of implementation clearly and, consequently, to examine the applicability level in terms of common technological classes.

The following three reference categories were identified during the case study discussion:

- Technological unit, an element identified within a grouping of functions necessary for obtaining environmental performances;
- Technological system, a structured set of technological units;

- Architectural organism, a unitary and structured set of technological systems.

The following application fields correspond to the reference macro-categories:

- Bearing structure;
- Vertical and horizontal envelopes;
- Internal partitions;
- External partitions.

The need to clearly define the reference categories and the application fields determines an approach aimed at further addressing construction problems of the building process. The systemic analysis of case studies is necessary for a comparative study to delineate the critical points and the potentials deriving from a possible implementation and transposition of the procedures in the realization of 1:1 scale architectural organisms. Each case study is examined through the definition of parameters related to the computational and manufacturing operation. The focus is centered on the computational paradigm, the methodology for form creation and subsequent discretization, identifying the variable and invariable geometric criteria that stipulate the space of design possibilities. As for the generative computational process, the analysis of digital manufacturing methodologies outlined the main traits of the technologies involved and the parameters that describe the tools and the materials used.

6.3.1 Structural Technological Systems

The survey conducted on the implementation of informed structural, technological systems is one of the lines of research that has given greater results with concrete applications in architecture. The prototypes realized through employed research programs involved the use of a wide range of materials, from wood to fibers, to plastic polymers, and of different digital manufacturing methods. The experiments that present a greater level of applicability are those that involve wood as a construction material if they are evaluated in terms of computational processes of structural optimization and experimentation on complex spatial configurations derived from the implementation of innovative digital manufacturing methods. An important investigation path is represented by structural wood technological systems starting from consolidated know-how, in terms of material properties and related construction technologies. This specific research path, involving digital computation and robotic fabrication, is particularly useful for all the implications concerning the exploration of new formal codes and parameters related to environmental sustainability, such as carbon footprint and embedded energy. The tests employed engineered elements, such as laminated and profiled panels, and natural, low-engineered materials, while

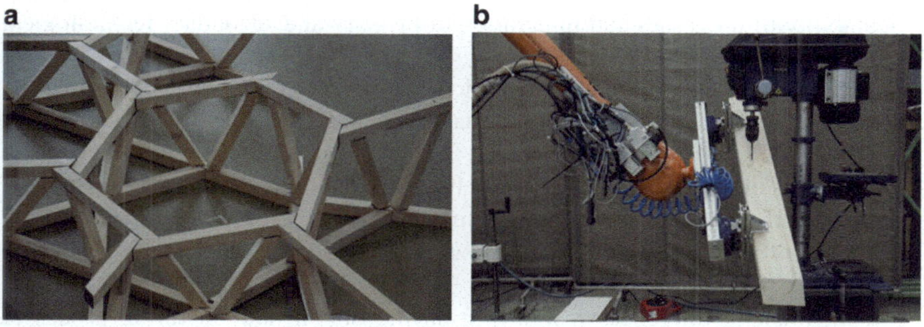

Fig. 6.1 **a** *Additive Robotic Fabrication of Complex Timber Structures*: prototype fabrication process. **b** *Additive Robotic Fabrication of Complex Timber Structures*: space frame scheme with expanded nodes. © Gramazio Kohler Research, ETH Zurich

digital exploration concerned structural and multi-objective maximization, computational procedures, as well as the study of principles of biomimesis[4] which could be abstracted and applied to the design of intricate and organic structures.

The digital manufacturing operations involved in the construction of wooden structural systems are manifold, and they include the following:

- Subtractive processes, for the realization of differentiated joints derived from complex and optimized spatial configurations;
- Additive procedures,[5] for the assembly of components using elaborate and informed spatial sequences;
- Combined, subtractive, and additive processes, in which the construction sequence involves machining the components before their assemblage.

Some of the wood structural systems examined tested the integration between different technological systems, main structure, secondary structure, insulation, finishing, and systems, demonstrating their actual applicability in the architectural field, validating the processes in terms of management of tolerances derived from innovative tools and, in some cases, homemade. Projects such as *Sequential Roof* (ETH) which is the roof structure of the *Arch_Tec_Lab* of the Institute of Technology in Architecture (ITA), as well as the entire production of the research path on additive processes (Fig. 6.1) for the construction of complex wooden structures of Gramazio Kohler Research, *Wood Chip Burn* of the AA, the Research Pavilion 2011 of ICD, basis for the building of the *Landesgartenschau Exhibition Hall* of 2014, proved the efficiency of computational procedures in the exploration of different design solutions and the concrete applicability of technological systems.

These experiences can be considered as proof of the concept for the following:

[4]Biomimesis is the study of the structure and function of biological systems as models for the design and engineering of materials and machines.

[5]The assembly of components can be assimilated to other additive processes.

- Scalability of processes and integration of primary and secondary technological systems;
- Potential of the morphogenetic operation and digital computation to enable the transfer of biological principles in virtual space, and subsequently in the physical one;
- Advantages deriving from genetic and heuristic maximization procedures in the design of structural performance systems;
- Potential of structural joints' customization for the construction of optimized structures;
- Management of tolerances resulting from the digital manufacturing processes used.

The construction of two research pavilions, *Digital Urban Orchard* and *Fusta Robòtica*, made at the IAAC in Barcelona, were a further test. The design of the two pavilions highlighted a further aspect regarding the applicability of structural wooden systems in architectural practice. The control of formal generation through the parametric process and the possibility of customization offered by robotic fabrication allow to overcome, or better expand, the concept of structure to that of an integrated system capable of aggregating primary and secondary structures, systems, and furnishings through specific morphological configurations. The studies concerning this topic present a high level of employment due to the direct involvement of industries stimulated by the sharing of know-how and the options offered by the combination of traditional materials, such as wood, and digital operations of performance simulation and digital manufacturing. Compared to the projects mentioned above, other design experiences have a more experimental vocation, in some cases mere speculations, which correspond to a low level of applicability in architectural practice, mainly in terms of scalability of procedures. Nevertheless, starting from the analysis of such projects, it is possible to extract theoretical concepts and future implementations that can be transferred to the design processes. One of these projects has to do with the investigation of soft systems, which contrasts with the hard systems discussed above, based on the adaptive relationship of a feedback loop between performance parameters and formal generation. The studies that combined responsive morphologies with the robots' introduction directly on the construction site clearly showed the potential of their application to structural, technological systems or performative envelopes at the architecture scale. A further research path concerns structural, technological systems that utilize composite materials, specifically carbon and glass fibers. The employment of new technological networks for structural applications was experimented through the construction of a series of prototypes, to benefit from the bidirectional workflow between formal generation and robot manufacturing. The research pavilions realized by the ICD (Fig. 6.2) on the subjects of Fibrous Tectonics, corroborate to concrete applicability of systems in the architectural and engineering field, even though limited to certain structural and spatial configurations. Furthermore, through the construction of the prototypes, it was possible to test an innovative production system that uses glass and carbon fibers without mold to build

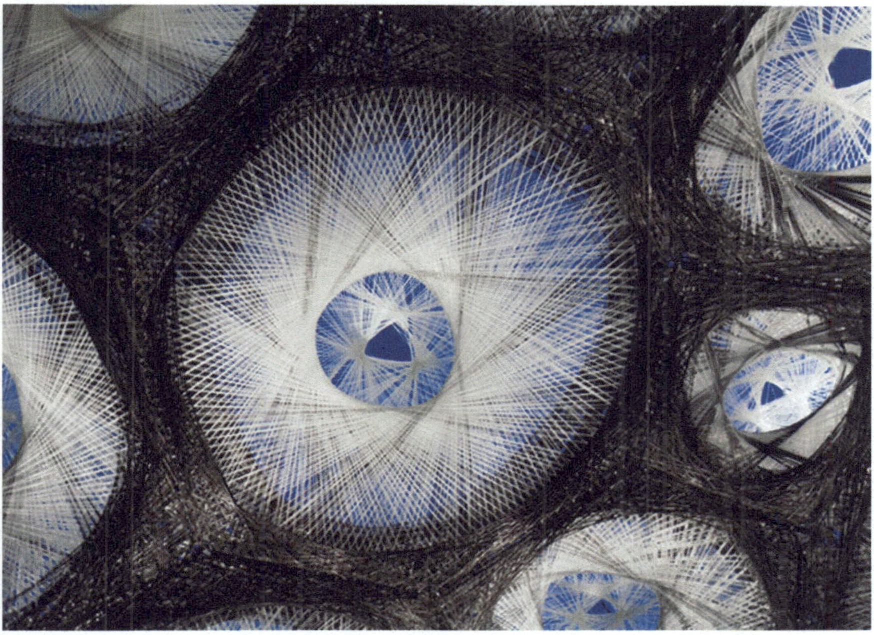

Fig. 6.2 ICD *Research Pavilion 2013/14*, materials composites for the realization of structural, technological units. © ICD/ITKE University of Stuttgart

complex and optimized shapes, with considerable savings in terms of fabrication times and costs.

The continuous experimentation overcame a series of limitations typical of the first design experiences conducted on the same topic. The development of communication protocols between computers and robots and the employment of other types of machines, such as drones, allowed increasing the size of components manufacturable in the laboratory, shifting from the construction of technological systems assembled on site, to an architectural organism that can be transported and installed later on. On the other hand, critical issues persist in the integration of the proposed structural systems with the other technological networks constituting an architectural organism (e.g., secondary structure, facilities, finishes). The last class of structural and technological systems analyzed uses metal materials as resistant elements. The Mesh Mold project of ETH (Fig. 6.3) is an example of how emerging technologies in the design process, from the design procedure to robotic manufacturing, allow to expand the range of design solutions and manage the formal differentiation, as well as customize the production process by designing customized tools. The efficiency of the operating method has guaranteed the possibility to carry out accurate tests on the material and structural system, first by utilizing polymeric components and then moving on to definitive prototyping with a metallic material. The wire mesh folded and welded onsite through a robotic manufacturing operation represents the

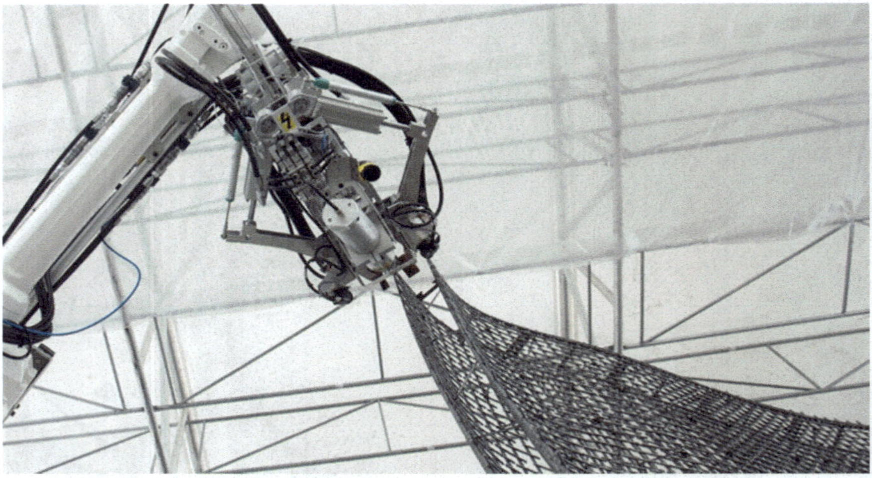

Fig. 6.3 *Mesh Mold Metal.* Structural metal mesh using a robotic arm and a customized tool. © Gramazio Kohler Research, ETH Zurich

structural system and the concrete casting formwork simultaneously. This kind of technology has a high level of applicability for the construction of free-form concrete structures and is linked to the use of robots directly on site.

6.3.2 Technological Systems for Vertical Envelopes

Given their performance characteristics and dimensional parameters, some of the technological systems analyzed can be applied as external envelopes. The data-driven process, performance optimization, and innovative digital manufacturing technologies, open up new investigation scenarios for research on opaque envelopes. The major innovation on the subject concerns the construction and technological operation: from external stratified and increasingly complex partitions that, layer-by-layer, enclose the main structure, the secondary structure, isolation, and coating, the research moves to integrated technological systems where formal generation is directly connected to the performance parameters that guide the design process. The primary studies on the topic can be traced back to two macro-categories, which are summarized as follows:

- Reinterpretation of units and technological systems commonly employed in the construction of external partitions through an innovation process;
- Design of new technological systems with digital manufacturing methods and polymeric materials to achieve informed and optimized spatial configurations.

As for wooden, structural, technological networks, the survey path on external envelopes involves all manufacturing methods accessible thanks to the generic nature of robots, while in terms of employed materials, the researches focus mainly on ceramic and plastics, with reference to the family of thermoplastic polymers in additive processes. One of the fundamental innovations is represented by the reinterpretation of traditional technological systems in clay by using new production technique while maintaining the technological unit, the brick. The computational process and the flexibility and efficiency of the industrial machine allow tailoring the final shape, to the scale of the technological system, as well as the morphology of the components, regarding the individual unit. Projects such as the Gantenbein Vineyard (Fig. 6.4), a customized façade created by Gramazio and Kohler in collaboration with the ETH, and the research prototypes conducted by the Robotic Design Group at Harvard University have shown the applicability of these technological systems to the architectural world.

The first project investigates the relationship between performative morphology and the robotic procedure of adaptive assembly that leads to the design and construction of an external envelope which meets all the performative requirements. The research of the Harvard Robotic Research Group examines the efficiency of subtractive processes to customize the individual components that are later assembled according to the traditional method for brick fabrication. The performances that underline the differentiation of form respond mainly to structural and energy aspects,

Fig. 6.4 Gantenbein Vineyard, Gramazio Kohler Research, customized façade system results of a robotic additive manufacturing process. © Gramazio Kohler Research, ETH Zurich

while the responsiveness of the technological system goes through its geometric optimization. Through this design methodology, a traditional building element, always linked to standardized production and construction processes, is reinterpreted through innovative design and manufacturing operations. Its form is differentiated based on performative criteria identified by the designer as well as being a guide to the data-driven design procedure. The second macroarea of investigation regards Additive Manufacturing (AM) for building envelopes with ceramic materials, especially thermoplastic polymers. 3D printing is a production method that represents a paradigm shift in components and technological systems' design and fabrication operation in terms of form customization, production processes, and materials utilized. The deposition of material, only where strictly necessary, is in accordance with the distribution paths of loads and structural forces. This is based on energy-environmental inputs linking the dimensional parameters of the shape, such as layer thickness, height, and correlation between solid and void, to performance, creating a bidirectional relationship between multi-objective optimization and digital fabrication. The studies conducted at IAAC on AM and clay for large-scale production, have demonstrated partial applicability of the technological systems proposed in architectural practice (Fig. 6.5). In reality, data on the durability of the proposed material system are lacking, as is lacking a proper analysis on structural behavior and structural joints that would guarantee a performative behavior and relate it to the complex association between interior and exterior (e.g., impermeability to air and water, resistance to atmospheric agents).

Fig. 6.5 IAAC *Terraperforma*, additive manufacturing and natural material for technological devices. © IAAC, Institute for Advanced Architecture

Regarding the innovation of the materials used, the combination of 3D printing and robotic fabrication tested, for the first time, plastic materials in the digital construction process of technological systems for the opaque envelopes.

The main researches on the subject are carried out at TU Delft and TU Eindhoven, and they can be summarized as follows:

- Robotic Additive Manufacturing to produce technological systems that can function as passive heating and cooling devices;
- 3D printing to manufacture active façade systems able to function as heat storage thanks to internal circulation of fluids (Fig. 6.6).

Both research paths work on a digital workflow for the design of informed technological systems founded on the integration between an optimized computational process and the Robotic Additive Manufacturing (RAM) technique. The relationship between geometric parameters, maximization, and production procedure is managed in the digital space in such a way as to guarantee the effective production of the components.

The results of the research on the matter have demonstrated the effectiveness and potential of the workflow in terms of the possibility to customize the shape based on

Fig. 6.6 TU Delft, 3D printing for active façade systems. © Michela Turrin

performative criteria. Despite this, the application of technological networks, both ceramic and plastic, on the architectural scale is subject to the resolution of specific critical issues, which can be summarized as follows:

- System of connection between technological units and design joints able to assure structural performance suited to regulatory standards;
- Development of eco-friendly materials with feature performance adapted to construction process' standards;
- Optimization of the production procedure to decrease fabrication time and, consequently, costs;
- Assessment of performances related to material's durability, static behavior, and tightness of the technological systems during post-production and installation phases.

6.3.2.1 Technological Systems for Internal Walls/Partitions

The technological systems for internal partitions represent an implementation field of absolute interest and with a high degree of applicability in architectural practice. As for external envelopes, the design of units and technological networks used as internal partitions, horizontal or vertical, benefits from the potential offered by the generative process and digital fabrication. The main focus is on the design of informed technological systems for the improvement of indoor comfort. In this regard, the use of the computational procedure makes it possible to link from generation to the analysis of performances, which refers mainly to the acoustic phenomena, while the geometric differentiation that derives from the process is managed due to robots and 3D printers.

In terms of the criteria regulating the acoustic comfort of spaces, it is possible to refer to the following parameters:

- Sound absorption;
- Attenuation of sound propagation;
- Soundproofing;
- Reverberation period.

The design of technological devices able to increase the acoustic performance of rooms for musical events, constitutes a field of specialization that has characterized the architectural production throughout history, even if through different operating methods. The innovative manufacturing technologies mentioned above have allowed to expand the range of design possibilities and, at the same time, introduce new materials that present physical traits and mechanical properties suited for the realization of this type of components. The design of responsive technological systems for indoor spaces can benefit from all the digital manufacturing techniques regarding subtractive and additive processes and opens to the experimentation of a wide range of traditional and new materials. The research conducted by ETH Zurich and TU Delft focuses on the definition of an algorithmic procedure for material system information based on

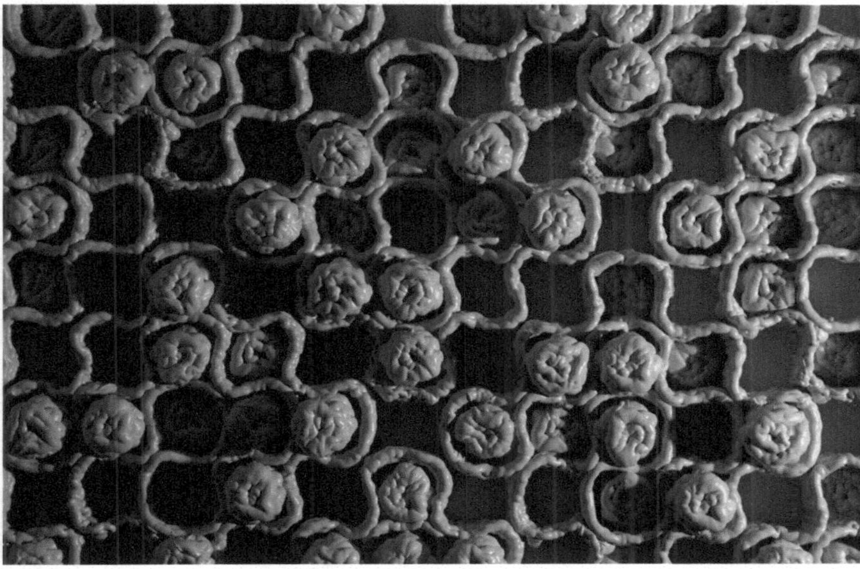

Fig. 6.7 ETH, *Acoustics*, technological system of internal partition in polyurethane foam. © Gramazio Kohler Research, ETH Zurich

the fabrication method implemented. The components utilized for the experimentation are several: from thermoplastic polymers to polyurethane foam, from subtractive operations to additive processes. In order to validate the digital procedure, after the installation of the components, the studies operated the evaluation of performance (Fig. 6.7). The prototypes realized and the concrete employments in architectural projects, show how the adopted methodology is transferrable to architectural practice. The applicability of the systems depends on the scale of the single technological units, which allows an easy computation and production process, as well as a simplification of the design of the connection between the various elements. At the current state of research, the main implementations on the topic concern the digital communication protocol between software (design and simulation) and hardware (fabrication), and the simplification of acoustic phenomena simulation in a digital environment; the latter serves as a basis for the optimization procedure of components.

6.3.2.2 Technological Systems for External Partitions

Similar to technological devices utilized to increase acoustic performance in indoor spaces, the design and construction of technological systems for external partitions, such as solar shading, represents a research field characterized by a high level of applicability in architectural practice. Shading systems, real buildings interfaces between the complex internal conditions and the external environment, are fundamental to

Fig. 6.8 Ceramic shading system, Harvard Robotic Group+Ascer, shielding ceramic system. © Martin Bechthold

guarantee indoor comfort conditions. The principles determining thermal comfort conditions, visual, acoustic, and air quality, which may be related to the presence of shading systems can be summarized as follows:

- Control of natural ventilation;
- Absorption and release of energy;
- Shading (Fig. 6.8);
- Control of natural lighting;
- Control of glare phenomenon, glare protection;
- Sound absorption;
- Attenuation of sound propagation;
- Soundproofing.

The potential of the computational process during the entire project development path, from the preliminary design phase to the fabrication of the components, is reflected in two main advantages. The optimization of geometric patterns based on performance parameters (a result of the digital simulation of the phenomena described above); and the definition of the dimensional criteria for the single technological units founded on the manufacturing methodology and material properties. The main research paths on the topic are as follows:

- Exploration of responsive patterns and determination of relative geometric parameters concerning the digital fabrication process;
- Development of digital workflows integrating computational maximization procedures with Virtual Reality (VR) 32 techniques and with digital empolyments that

allow the user to customize the form based on parameters according to the specific project application;
- Study of secondary structure and structural joints among the various technological units;
- Creation of materials with technical and performance characteristics suitable for outdoor applications, especially for the thermoplastic polymers in 3D printing.

Experimentation on external partitions, additive processes, and 3D printing, represents an innovative and efficient production method for the prototyping of components, given the size of the basic technological unit. The use of a relatively quick and cheap prototyping method, realizable with a common desktop 3D printer, enables to accelerate the experimentation and test the structural behavior of technological systems and utilize environmentally sustainable materials. Furthermore, the use of new manufacturing technologies allows testing with formal evolutionary codes, exploring complex and organic design solutions whose materialization is made possible by the innovative technologies utilized in the procedure. The main critical issues concern the performance of the materials used in terms of compliance to the standards proposed by the legislation on fire resistance and atmospheric agents, and the technological joints between different units, and between the external partitioning system and the main façade structure.

6.4 Paradigm Shift in Design and Construction Processes

As highlighted in the examination of the projects, the problematic points of greatest importance can be identified in all the technological classes of reference investigated and concern the computational process, the material system, and the technological and constructive aspects. In the computational procedure, the first problem treats the need for sector operators with integrated skills to manage the entire design operation. The data-driven strategy envisages the integration of formal generation and digital manufacturing process within the same workflow, and this implies a greater management complexity compared to common design processes. A further critical trait is a complexity deriving from multi-objective computational optimization paths and the consequent discretization of the parametric model necessary to limit calculation times and to guarantee readability and usability of the data obtained as a result of the maximization. Compared to the production and manufacturing phase, the operational methodology adopted, based on the customization of tools and processes, involves the design of complex communication protocols between digital model and robot, as well as special end effectors. The management of the procedure defined as cyber-physical making requires the possession of specialized skills on the border between systems mechatronics, electronic engineering, and computer science. Such skills are not easily available among the sector's operators. The paradigm shift in progress leads to a further reflection on the materials used in the production processes in terms of their physical characteristics and suitable mechanical properties to employ them in the architectural practice. The analysis shows that one of the major problems concerns the technological-constructive operation associated with the implementation of the components. The management of tolerances, the result of non-engineered manufacturing processes involving customized tools, and at times homemade, requires the design of communication protocols based on the feedback loop relationship and needs to adapt the fabrication procedure to the digital model and vice versa. Tolerance becomes an important design theme together with the scalability of operations and study of the structural connections between technological units and integration with the other constructive elements constituting an architectural organism. From the analysis carried out, it is clear how the definitive introduction of the proposed systems in architectural practice needs to address the resolution of technical problems and, above all, come to terms with the rethinking of design and construction processes' complexity. The first step reviews the centrality of the production procedure that becomes an integral part of the design operation, starting from the preliminary and ideational stage, helping to determine the limits of the formal creation process and, at the same time, ensuring the feasibility of the components. In order to do this, it is necessary to have a structured knowledge of digital manufacturing processes and production methods made available by the use of generic machines, such as industrial robots. The indispensable knowledge regards the technical specifications of the instruments, the organization of the work area (robot cell) and the performance traits of the tools to perform a specific production process. Furthermore, the customization of the fabrication implies a series of operations related to the calibration of the same for a correct management of tolerances and to guarantee the perfect correspondence between the physical environment and the digital space. The second aspect concerns the development of a new material sensitivity aimed at overcoming the consolidated relationship between material and production process that characterizes industrial operations. The material acquires a high specific weight in the preliminary phase of the design and its correct management depends on the success of the operation. The development of a material sensitivity and the conferral of digital intelligence is fundamental in the process of formal generation to investigate its performance principles through a feedback loop correlation of continuous interaction. The digital procedure is informed by the physical and mechanical properties of the materials and the constraints derived from the manufacturing method utilized, which consequently expands the range of design possibilities and introduces a new performative layer. The conjugation of the two terms apparently opposed as digital, representation of the virtual environment, and material, as something tangible and concrete, finds complete expression thanks to the computational-algorithmic approach. The customization of the production process and the renewed materiality imply a redefinition of the constructive logics as a combination of the performance of the material, expression of its technical characteristics and its mechanical properties, and new manufacturing methods.

References

1. Bechthold M (2012) Performalism or performance based-design. In: Grobman Y, Neumann E (eds) Performalism: form and performance in digital architecture. Routledge, Abingdon
2. Cui C, Ohmori H, Sasaki M (2003) Computational morphogenesis of 3D structures by extended eso method. J Int Assoc Shell Spat Struct 4:51–61
3. Deutsch R (2015) Data-driven design and construction: 25 strategies for capturing, analyzing and applying building data. Wiley, Hoboken
4. Grobman Y, Neuman E (2012) Performalism: form and performance in digital architecture. Routledge, London and New York
5. Kolarevic B (2003) Architecture in the digital age. Design and manufacturing. Spon Press, New York
6. Kolarevic B, Malkawi A (2005) Performative architecture: beyond instrumentality. Spon Press, New York
7. Leatherbarrow D (2009) Architecture oriented otherwise. Princeton Architectural Press, New York
8. Luebkeman C (2003) Performance-based design. In: Kolarevic B (ed) Architecture in the digital age-design and manufacturing. Spon Press, New York, pp 372–388
9. Oxman R (2009) Performative design: A performance-based model of digital architectural design. Environ Plan 36:1026–1037
10. Oxman R (2015) MFD: Material-Fabrication-Design: a classification of models from prototyping to design. In: Proceedings of the international association for shell and spatial structures (IASS) Symposium 2015, Amsterdam
11. Proving Ground (2015) Using data in your design process. http://provingground.io/2015/08/27/using-data-in-your-design-process/. Accessed 27 August 2016
12. Tedeschi A, Wirz F (2014) AAD-Algorithms-Aided Design: parametric strategies using Grasshopper. Le Penseur publisher, Brienza

The manufacturer's authorised representative in the EU is Springer Nature Customer Service Centre GmbH, Europaplatz 3, 69115 Heidelberg, Germany. If you have any concerns regarding our products, please contact ProductSafety@springernature.com

Printed and bound by CPI Group (UK) Ltd, Croydon, CR0 4YY

30/09/2026

02225574-0018